Systematic Analysis of Bipolar and MOS Transistors

The Artech House Materials Science Library

Amorphous and Microcrystalline Semiconductor Devices, Volume I: Optoelectronic Devices, Jerzy Kanicki, ed.

Amorphous and Microcrystalline Semiconductor Devices, Volume II: Materials and Device Physics, Jerzy Kanicki, ed.

Electrical and Magnetic Properties of Materials, Philippe Robert

High-Speed Digital IC Technologies, Marc Rocchi, ed.

Indium Phosphide and Related Materials: Processing, Technology, and Devices, Avishay Katz, ed.

Introduction to Semiconductor Device Yield Modeling, Albert V. Ferris-Prabhu

Materials Handbook for Hybrid Microelectronics, J.A. King, ed.

Microelectronic Reliability, Volume I: Reliability, Test, and Diagnostics, Edward B. Hakim, ed.

Microelectronic Reliability, Volume II: Integrity Assessment and Assurance, Emiliano Pollino, ed.

Modern GaAs Processing Techniques, Ralph Williams

Semiconductor Quantum Wells and Superlattices for Long-Wavelength Infrared Detectors, M. O. Manasreh, ed.

Vacuum Mechatronics, Gerado Beni, Susan Hackwood, et al.

VLSI Metallization: Physics and Technologies, Krishna Shenai, ed.

Systematic Analysis of Bipolar and MOS Transistors

Uğur Çilingiroğlu

Artech House
Boston • London

Library of Congress Cataloging-in-Publication Data

Uğur Çilingiroğlu
Systematic Analysis of Bipolar and MOS Transistors/Uğur Çilingiroğlu
Includes bibliographical references and index.
ISBN 0-89006-625-6
1. Bipolar transistors—mathematical models. 2. Metal oxide semiconductors—mathematical models. I. Title

TK7871.96.B55C55 1993 92-43084
621.3815'28—dc20 CIP

British Library Cataloguing in Publication Data

Uğur Çilingiroğlu
Systematic Analysis of Bipolar and MOS Transistors
I. Title
621.3848

ISBN 0-89006-625-6

685 Canton Street
Norwood, MA 02062

International Standard Book Number: 0-89006-625-6
Library of Congress Catalog Card Number: 92-43084

10 9 8 7 6 5 4 3 2 1

To Idil, Emre and Ioli

who have suffered a lot

Contents

Preface

This book is intended to guide students of microelectronics along the analytical path of semiconductor device modeling. A working knowledge of this path enables the VLSI designer to use device models intelligently, to customize them as needed, and even to develop new ones. Those microelectronics students trained to understand that device modeling is a peripheral task to be undertaken solely by the solid-state experts are destined to be passive users of prescription models only. The experience of the "populist" VLSI movement of the early 1980s has clearly shown this to be true.

Semiconductor device analysis has a very general mathematical basis comprising five fundamental equations and a set of supplementary ones. These equations summarize the relevant constraints originating from solid-state physics, statistical mechanics, and the theory of electromagnetism. Their solution with appropriate boundary conditions eventually leads to device models that interrelate electrical port variables in terms of structural parameters. Except for a few cases, however, exact solutions do not exist, a fact that turns device modeling into an art of approximations. Performing it efficiently on a given problem calls for the ability to discover the most appropriate associations along the deductive path connecting the generality of fundamental equations to the singularity of device structures. This skill is the single most important attribute in device modeling, and, therefore, must be among the primary objectives of microelectronics education. A course structure intended to fulfill this objective should guide the student without distraction through the seemingly fruitless front end of deduction all the way up to the end-product models. This must be accomplished in a period of ten to fifteen weeks, and in the end the entire path of deduction left behind should still remain visible. These constraints necessitate a highly streamlined and closely integrated course structure. This book reflects my conception of such a structure.

Streamlining has been implemented in three areas. First of all, the physical origins and technological aspects of device engineering have been excluded. This is an area addressed by a large number of excellent books. Any interested student can access this area through the current literature. Secondly, device coverage has been reduced to include only the conventional BJTs and MOSFETs. Once the student has acquired the analytical skills as intended, he or she can easily move from the upper end into those problems involving less conventional devices. The unique aspects of such devices are also well documented in easily accessible reference books. The third and probably most critical area of streamlining is the path connecting the two ends. The traditional emphasis on modeling *pn*-junctions and MOS capacitors as pedagogical intermediaries between fundamental concepts and transistors has been abandoned. Instead, transistor structuring has been lowered to a more primitive level; neutral regions and space-charge regions are treated as the main structural invariants. Thanks to this approach, the sequence of deduction remains short and unified. In the mean time, great care has been exercised to maintain a high level of integrity throughout the text. At the end of each chapter, the reader is urged to test his or her level of accomplishment by engaging in a number of exercises, some of which have been extracted from research articles published in the past few years. Hopefully, these exercises will thus familiarize the reader with the recent literature on device modeling.

This book has been written primarily for graduate students of microelectronics. Having been already exposed to solid-state devices on least at an elementary level, a graduate student can easily appreciate the message given and endure the discipline demanded. However, senior undergraduate students and practicing engineers should also benefit from the material presented.

This book is based on the lecture notes I have been using for many years now at Texas A&M University and Istanbul Technical University. My students at both locations have shaped up the evolution of these notes. Without the encouragement of my dear friend and colleague Donald Parker, however, I would have never undertaken the hardship of turning the raw material into a book. Fortunately, I have had on my side the assistance of Banu Pamir and Hakan Ozdemir as well as the understanding of Pamela Ahl to endure this hardship. I thank them all.

U. Çilingiroğlu
Princes Islands
March 1, 1993

Chapter 1
Background

1.1 FUNDAMENTAL CONCEPTS AND EQUATIONS

In this section, we briefly introduce the basic properties of semiconductor material pertaining to device operation. After reviewing the constituents of a semiconductor crystal, we will derive five fundamental equations to represent the main physical phenomena involving these constituents. Also presented are several auxiliary relationships and a number of analytical tools. This section will thus translate semiconductor physics into the domain of device analysis and modeling. This translation is intended to be as concise as possible. Those who wish to gain a deeper insight into the physical roots of the subject matter are referred to Pierret [1,2], Warner and Grung [3], Shur [4], Ferendeci [5], and Sze [6], which are excellent resources for additional information for other sections of this chapter as well.

1.1.1 Constituents of a Semiconductor Crystal

Bipolar junction transistors (BJTs) and metal-oxide semiconductor transistors (MOSFETs) are built with crystalline semiconductor material, most commonly, with silicon. The main constituents of a silicon crystal are illustrated in Figure 1.1 on a simplified two-dimensional diagram. We observe the following features:

1. Most of the lattice sites are occupied by *silicon atoms*. These sites are neutral because a core charge of $+4q$ (contributed by the nucleus and inner-shell electrons) is neutralized by four *valence electrons*. Note that q stands for the magnitude of the electronic charge, that is, $q = 1.6 \times 10^{-19}$ C. Each valence electron is shared by two neighboring silicon atoms. This interaction, called the *valence bond,* is the essence of the crystalline form. It holds the atoms together and thus keeps the material in a solid state.

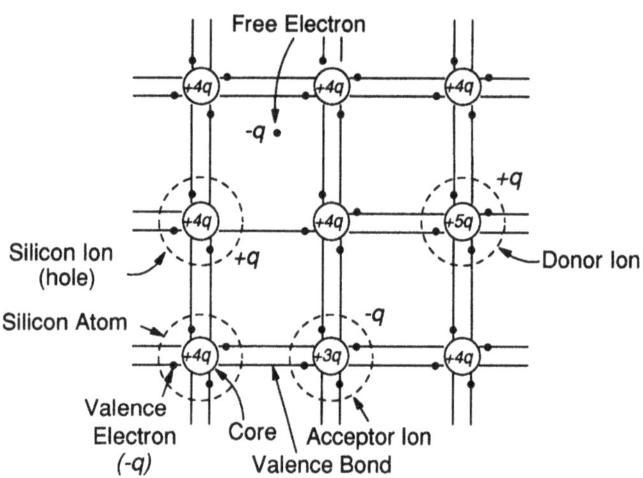

Figure 1.1 The main constituents of a silicon crystal pertaining to solid-state electronics. The variable q depicts electronic charge, that is, $q = 1.6 \times 10^{-19}$ C.

2. Some lattice sites are occupied by *donor ions.* These are introduced to the silicon crystal as five-valence atoms (such as phosphorus and arsenic) in a fabrication process called *doping* [7]. Such a *dopant atom* settles into a lattice site by forming four valence bonds with it neighbors. Its fifth valence electron, being unable to form a valence bond, leaves the atom and is donated to the crystal. As a result, the dopant atom is ionized with a net charge of $+q$ because a total of $+5q$ core charge is now surrounded by a total of only $-4q$ valence charge. This is why we call it a *donor ion.*

3. Some lattice sites are occupied by *acceptor ions.* These are introduced by doping with three-valence atoms (typically, boron). Each forms four valence bonds using its own three-valence electrons plus one electron taken away from the crystal. Since it accepts an electron, it turns into an acceptor ion charged to $-q$.

4. At some lattice sites, a valence bond may be missing. This is possible because a valence electron, having acquired sufficient energy from the thermal reservoir of the crystal or from other energy imparting sources, can shake itself loose from the host atom. What is left behind is a *silicon ion* charged to $+q$. These ions are called *holes.* As depicted in Figure 1.2, a missing bond in atom A can be easily filled in by a valence electron transferring from a nearby atom B. This process of electron transfer neutralizes atom A but, obviously, leaves a vacant bond (a hole) in atom B. We therefore conclude that the presence of holes in a semiconductor enables valence electrons to become mobile.

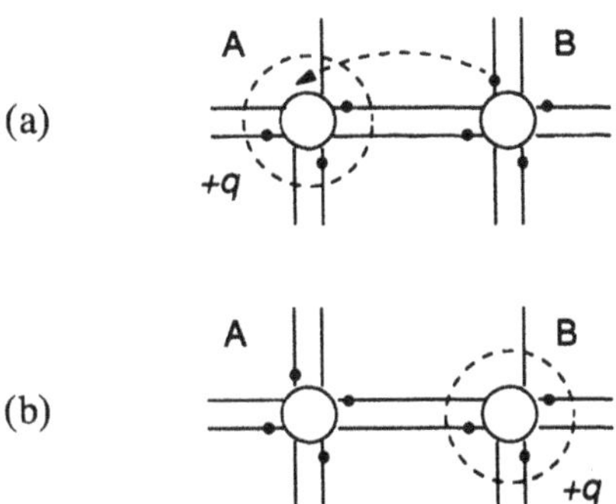

Figure 1.2 (a) A valence electron of atom B transfers to a vacant valence bond site in atom A. (b) The transfer results in hole movement from atom A to atom B.

5. The silicon crystal also contains *free electrons,* which are too energetic to belong to any particular site. These electrons are naturally mobile. The distinction between the free-electron and valence-electron energies can be best visualized with the aid of the so-called *energy-band diagram,* which, as shown in Figure 1.3, is an illustration of the levels of energy electrons can assume in a given semiconductor. The energy levels pertaining to semiconductor action are seen to be grouped into three distinct bands. The band that comprises the levels below E_V is called the *valence band.* This is the band of valence-electron energies. In this band, each occupied level corresponds to a valence electron, whereas an unoccupied level signifies a hole. The band lying above E_C comprises the energy levels assumable by free electrons and is called the *conduction band.* These two bands are separated by a third band of width $E_g \equiv E_C - E_V$. This band is known as the *bandgap* because in an ideally pure semiconductor no energy level can exist in it. In reality, however, the bandgap contains levels created by crystal imperfections. Such imperfections are unintentionally and inevitably introduced to the crystal during fabrication in the form of atomic contaminants and crystal defects and are commonly known as *traps* because they can temporarily capture and immobilize free electrons and valence electrons.

Having summarized the main constituents of a semiconductor crystal, and knowing that all devices contain the very same entities, we might ask what makes a device distinguishable from other devices. The answer is twofold: (1) uniqueness of the geometry of the device, that is, the location of the boundaries, and (2) uniqueness of the doping profiles inside this geometry. Both features are fixed during device fabrication and remain invariant thereafter. In device operation the electric field and current at any location of the crystal are determined collectively

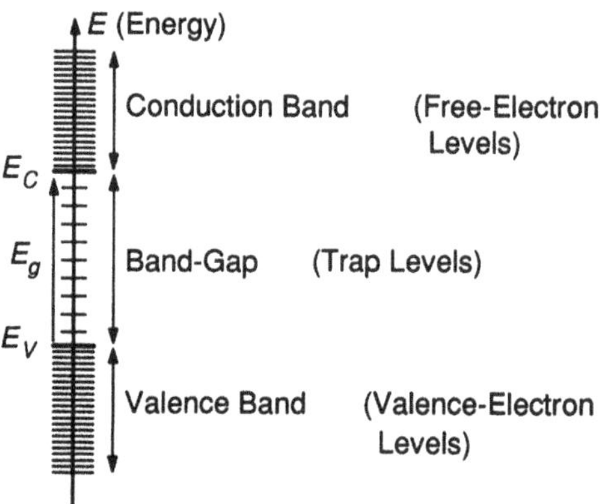

Figure 1.3 The energy-band diagram showing the three bands of electron energy levels, where E_C and E_V denote the lowest and highest levels in the conduction and valence bands, respectively, and E_g is the bandgap energy, for example, $E_g = 1.12$ eV in silicon at 300K.

by the dopant profile and externally determined boundary conditions in accordance with certain natural laws that govern the creation, disappearance, and transportation of the mobile constituents of the crystal. These laws are represented by a set of five equations: Poisson's equation, two current density equations, and two continuity equations. These fundamental equations are valid at any location of any device at any time. Their particular solution for the specific dopant profile and boundary conditions of a given device uniquely defines the state of the device. Whether we are interested in analyzing a device, for example, finding the response of internal variables to a given excitation, or in modeling a device, for example, finding a relationship between terminal variables in terms of structural parameters, the central problem will be the same: finding the particular solution of the fundamental equations! In the following subsections, we introduce each of these equations and give a brief description of the underlying physical phenomenon.

1.1.2 Poisson's Equation

Poisson's equation states that, at a given point of a dielectric material, the divergence of the displacement vector **D** equals the volumetric charge density ρ, that is,

$$\nabla \mathbf{D} = \rho \tag{1.1}$$

Also note that **D** is related to the electric field $\mathscr{E}$ by

$$\mathbf{D} = \epsilon\mathscr{E} \tag{1.2}$$

where ϵ is the dielectric constant. For a homogeneous dielectric, in which ϵ is independent of position, Poisson's equation can be written from (1.1) and (1.2) as

$$\nabla\mathscr{E} = \frac{\rho}{\epsilon}$$

whose one-dimensional form is

$$\frac{d\mathscr{E}}{dx} = \frac{\rho}{\epsilon} \tag{1.3}$$

According to Figure 1.1, the charged entities in a semiconductor are holes (charged to $+q$), free electrons (charged to $-q$), donor ions (charged to $+q$), and acceptor ions (charged to $-q$). Therefore, the volumetric charge density in a semiconductor can be generally expressed as

$$\rho = q(p - n + N_D - N_A) \tag{1.4}$$

where p, n, N_D, and N_A denote the volumetric concentrations of holes, free electrons, donor ions, and acceptor ions, respectively. Assuming a homogeneous semiconductor, we obtain from (1.3) and (1.4) the equation

$$\frac{d\mathscr{E}}{dx} = \frac{q}{\epsilon}(p - n + N) \tag{1.5}$$

where $N \equiv N_D - N_A$ is the *net doping concentration.* This is the most popular form of Poisson's equation in device analysis. Note that the unit of preference is centimeters for x; inverse cubed centimeters for p, n, N_D, and N_A; and volts per centimeter for $\mathscr{E}$. The value of ϵ for silicon is 1.04×10^{-12} F/cm.

Among the variables appearing in Poisson's equation, $\mathscr{E}$, p, and n are generally functions of time and position, while N_D and N_A can be functions of position but not of time because the distribution of dopants is fixed during device fabrication.

Poisson's equation is of enormous utility in device analysis. Most typically, it is used for determining the field profile in a semiconductor region of known charge density. Once the field profile has been obtained, one can also determine the electrostatic potential profile using

$$\mathscr{E} = -\frac{d\psi}{dx} \tag{1.6}$$

where ψ denotes the electrostatic potential, which is measured in volts. For a multidimensional field, (1.6) is expressed as

$$\mathscr{E} = -\nabla\psi$$

1.1.3 Current Density Equations

Physical Origins

The transport of electrical charge in a given direction across a given surface results in a current. As we discussed previously, the two types of transportable charged entities in a semiconductor are free electrons and valence electrons. An important distinction between the two is that, while all free electrons can participate in current conduction without any restriction, the number of valence electrons that can do so is limited by the number of available holes because a valence electron needs a hole in which to move. For this reason, the current that is actually conducted by valence electrons is determined, in effect, by the population of holes. Therefore, we conceive of the $+q$ charged holes as the effective carriers of the valence-electron current. For those of us who need more justification before accepting this idea, it might be helpful to note from Figure 1.2 the equivalency of the currents carried by a $-q$ charged valence electron moving to the left and a $+q$ charged hole moving to the right. Now that we have identified free electrons and holes as the current carriers of a semiconductor, we are ready to consider the mechanics of their motion and, hence, their current.

Carrier motion in a semiconductor has the following three key properties, which are illustrated in Figure 1.4:

1. The semiconductor crystal is packed with atoms and ionized dopants. Carriers, while propagating through the crystal, frequently collide with these immobile entities. A colliding carrier transfers energy to the crystal.
2. A carrier recovers from a collision by absorbing thermal energy from the crystal and starts moving in a *random* direction.
3. A carrier, while in flight between two consecutive collisions, is subject to a force if an electric field exists. This force adds a definite (nonrandom) acceleration component to the random thermal motion. Consequently, the trajectory of the carrier motion is bent in the same direction as the field for a hole and in the opposite direction for a free electron.

Considering the fact that the carrier motion just summarized is actually a three-dimensional motion, one can easily imagine the complexity involved in its

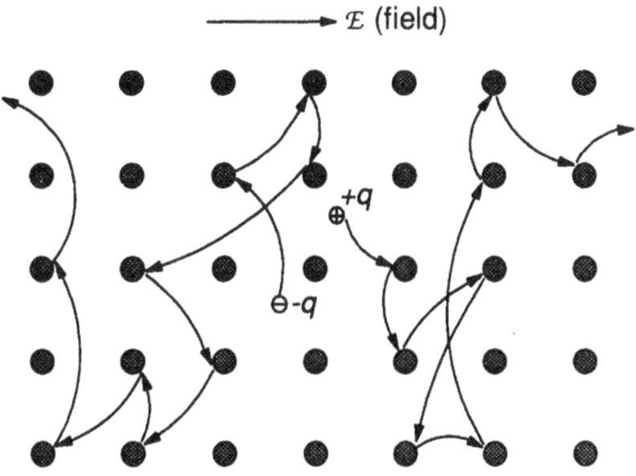

Figure 1.4 A schematic illustration of the free-electron and hole motions in response to an electric field $\mathscr{E}$. Arrows indicate the trajectories.

analytical description. We will now take a shortcut, and construct a one-dimensional model for the motion, which will, nevertheless, accurately describe the resulting current.

Mathematical Derivation

First, consider the holes depicted in Figure 1.5, and suppose that all have just recovered from a collision at time $t = 0$, and that their next collision is due at $t = \tau$. Assuming a one-dimensional motion along x, we expect one-half of the hole population to be right-bound and the other half to be left-bound because the direction along which a carrier exits a collision is random, as explained previously. Now, for the period between $t = 0$ and $t = \tau$, suppose we observe a surface S located at, say, $x = 0$. We expect to see all the right-bound holes that are initially located within a sufficiently close distance l_1 to the left of the surface (hole B, for example) to cross in free flight to the right during the observation period. Meanwhile, all the left-bound holes that are initially located within a distance l_2 to the right of the surface (hole B$'$, for example) will cross to the left. The difference between these two opposing hole flows corresponds to a hole current whose density J_p per unit surface area can be expressed as

$$J_p = \frac{1}{\tau}\left[\frac{q}{2}\int_{-l_1}^{0} p(x)\,dx - \frac{q}{2}\int_{0}^{l_2} p(x)\,dx\right]$$

The first and second terms inside the brackets correspond to the hole charge densities per unit area crossing to the right and left, respectively. By representing

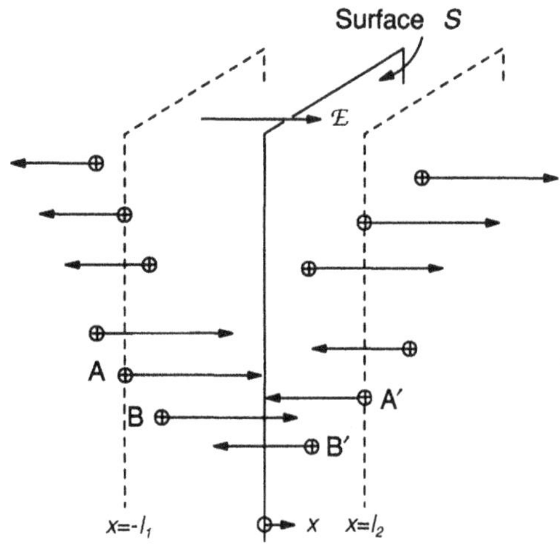

Figure 1.5 One-dimensional hole motion. The circles represent holes at $t = 0$ when the motion starts with the random thermal velocity v_{th}. The right-bound and left-bound hole populations are equal in size because the direction of the motion is random. If $\mathscr{E} > 0$, the right-bound holes are accelerated while the left-bound holes are decelerated. If $\mathscr{E} < 0$, the opposite situation prevails. The arrowheads indicate the final positions of the holes at $t = \tau$ when the free flight ends. Notice that only the right-bound holes with an initial location of $-l_1 \leq x \leq 0$ and the left-bound holes with an initial location of $0 \leq x \leq l_2$ can cross the surface S sometime during the observation period and thus can participate in current conduction across this surface.

$p(x)$ with the first two terms of its Taylor series, the above equation can be transformed into

$$J_p = \frac{q}{2\tau}(l_1 - l_2)p(0) - \frac{q}{4\tau}(l_1^2 + l_2^2)\frac{dp}{dx}\bigg|_{x=0} \tag{1.7}$$

We now need to determine l_1 and l_2. These, by definition, are the initial locations of the holes that are the latest to reach the designated surface at time $t = \tau$. Consider hole A, which starts its right-bound free flight at $x = -l_1$ with an initial (thermal) velocity v_{th}. Its motion for $t > 0$ is described by

$$\frac{d^2x}{dt^2} = \frac{q\mathscr{E}}{m_p}$$

where $q\mathscr{E}$ is the force due to an electric field, and m_p is the effective mass of the hole. Integrating this equation with the conditions

$$\left.\frac{dx}{dt}\right|_{t=0} = v_{th}, \qquad x(0) = -l_1$$

and then substituting $x = 0$ and $t = \tau$, we obtain

$$l_1 = (v_{th} + v_p)\tau \tag{1.8}$$

where

$$v_p \equiv \frac{q\tau}{2m_p}\mathscr{E} \tag{1.9}$$

is called the *drift velocity* of holes. A similar derivation involving the left-bound motion of a hole starting at $x = l_2$, such as hole A′, yields

$$l_2 = (v_{th} - v_p)\tau \tag{1.10}$$

Notice from (1.8) and (1.10) that a drift component is superimposed by the field on the otherwise random thermal velocity v_{th}. Unlike v_{th}, this component has the same definite direction for the left-bound and right-bound holes. This is why the average velocity in a right-bound motion, $v_{th} + v_p$, is different from its left-bound counterpart, $v_{th} - v_p$.

Substituting (1.8) and (1.10) into (1.7), and assuming a weak electric field, (hence, $|v_p| << |v_{th}|$), we finally obtain the *current density equation* for holes:

$$J_p = qpv_p - qD_p\frac{dp}{dx} \qquad [\text{A/cm}^2] \tag{1.11}$$

where

$$D_p \equiv \frac{\tau}{2}v_{th}^2 \qquad [\text{cm}^2/\text{s}] \tag{1.12}$$

is the so-called *hole diffusivity,* which can be regarded as a structural parameter.

Substituting (1.9) for v_p in (1.11), we obtain a more popular form of the hole current density equation:

$$J_p = q\mu_p p\mathscr{E} - qD_p\frac{dp}{dx} \tag{1.13}$$

where

$$\mu_p \equiv \frac{q\tau}{2m_p} \qquad [\text{cm}^2/\text{V}\cdot\text{s}] \tag{1.14}$$

is known as *hole mobility*. Notice from (1.9) and (1.14) the following relationship between drift velocity, mobility, and field:

$$v_p = \mu_p \mathscr{E} \tag{1.15}$$

The mobility itself is related to diffusivity through the so-called *Einstein relationship*. To derive this relationship, note that the average kinetic energy of a hole equals $(1/2)kT$, where $k = 8.614 \times 10^{-5}$ eV/K is the Boltzmann constant, and T is the absolute temperature.[1] Therefore, we can write

$$\frac{1}{2} m_p v_{th}^2 = \frac{1}{2} kT$$

which yields $v_{th}^2 = kT/m_p$. Substituting the latter into (1.12) for v_{th}^2, and comparing the resulting equation with (1.14), we obtain the Einstein relationship:

$$D_p = \frac{kT}{q} \mu_p \tag{1.16}$$

Also note that kT/q is called *thermal voltage* and has a value of 25.8 mV for the standard room temperature $T = 300\text{K}$.

By replacing the holes with free electrons in the foregoing analysis, we obtain the current-density equation for free electrons. All one needs to do is to replace q and m_p with $-q$ and m_n, respectively. Hence,

$$(1.11) \Rightarrow J_n = -qnv_n + qD_n \frac{dn}{dx} \tag{1.17}$$

$$(1.13) \Rightarrow J_n = q\mu_n n \mathscr{E} + qD_n \frac{dn}{dx} \tag{1.18}$$

where

$$(1.15) \Rightarrow v_n = -\mu_n \mathscr{E} \tag{1.19}$$

$$(1.14) \Rightarrow \mu_n \equiv \frac{q\tau}{2m_n} \tag{1.20}$$

$$(1.16) \Rightarrow D_n = \frac{kT}{q} \mu_n \tag{1.21}$$

[1] In actual three-dimensional motion, the energy equals $(3/2)kT$.

The current density equations, (1.11) and (1.17), or (1.13) and (1.18), are among the five fundamental equations of semiconductor device analysis. Their multidimensional forms are as follows:

$$\mathbf{J}_p = qp\mathbf{v}_p - qD_p\nabla p$$
$$\mathbf{J}_n = -qn\mathbf{v}_n + qD_n\nabla n$$

or

$$\mathbf{J}_p = q\mu_p p\mathscr{E} - qD_p\nabla p$$
$$\mathbf{J}_n = q\mu_n n\mathscr{E} + qD_n\nabla n$$

Obviously, the total current density in a semiconductor is the sum of the hole and free-electron currents, that is,

$$J = J_p + J_n \tag{1.22}$$

or, in a multidimensional form,

$$\mathbf{J} = \mathbf{J}_p + \mathbf{J}_n$$

The current density equations indicate two distinct current components for each type of carrier. Namely, a *drift* component, which is represented by the first term on the right-hand side of these equations, and a *diffusion* component, which is represented by the second term.

The drift current is due to a nonrandom displacement of carriers by an electric field in between consecutive collisions. It is obvious from (1.13) and (1.18) that this current flows in the same direction as the electric field regardless of the type of carrier involved. However, the direction of the underlying drift motion is opposite that of the field in the case of free electrons [see (1.19)] but is the same as the field in the case of holes [see (1.15)]. This is due to the opposing charge polarities of the two types of carriers.

As indicated by the current density equations, a diffusion current flows wherever the carrier concentration varies with position. Indeed, if the carrier concentration on one side of a surface is higher than that of the other side, more carriers are expected to cross, in a given time, from the high-concentration side to the low-concentration side in comparison with those crossing in the opposite direction. The reason for this is the randomness of the thermal motion. Obviously, the carriers diffuse in the direction of the decreasing concentration. The resulting current, on the other hand, is in the direction of increasing concentration in the case of the negatively charged free-electrons [see (1.17) or (1.18)], and is in the direction of decreasing concentration in the case of the positively charged holes [see (1.11) or (1.13)].

Mobility

In the two current density equations, the effect of carrier scattering is represented by the parameter of carrier mobility. We will now examine this important parameter in some detail.

First of all, free electrons and holes have different mobilities because the effective masses of these carriers, m_n and m_p, are different [see (1.14) and (1.20)]. Second, we expect the mobility to be proportional to the average time τ a carrier spends in free flight between two consecutive collisions [again see (1.14) and (1.20)]. Obviously, τ decreases with the density of the scattering centers. For this reason, both μ_n and μ_p are decreasing functions of the total doping concentration, as shown in Figure 1.6. It is important to note that the data in this figure are not applicable to the case of carrier transport in close proximity to a crystal surface where the presence of additional scattering centers further reduces the mobility. We will return to the topic of surface mobility when we study the MOSFET in which the transportation of carriers is confined to a narrow region neighboring a surface. The BJT, however, is a bulk conduction device to which the data of Figure 1.6 are directly applicable.

The mobility is a decreasing function of temperature as well. The dependence is more pronounced in a lightly doped material, in which scattering with the thermally vibrating atoms is predominant.

Judging from (1.14) and (1.20), the mobility appears to be independent of the electric field. This is indeed the case when the field is weak. In strong fields, however, the mobility becomes a decreasing function of the field, and consequently, the linear relationship between the drift velocity and field collapses, as depicted in Figure 1.7. Ultimately, the drift velocity saturates at about 10^7cm/s for both types of carriers in silicon.

1.1.4 Continuity Equations

We pointed out in Section 1.1.2 that the two carrier concentrations p and n could vary with time; but we did not discuss the causes of such temporal variations. What are the possible causes?

First, consider the hole population in a volume of semiconductor, and suppose the hole current entering the volume on one side is not equal to the hole current exiting on the other side; that is, J_p has a spatial variation. Obviously, the number of holes brought into the volume per unit time will be different from what is taken out in the same time. This will definitely lead to a temporal variation in the population of holes. Similarly, a spatial variation in J_n will result in a temporal variation of the free-electron population of the volume. This effect is perfectly analogous to the temporal variation in the population of a town of different im-

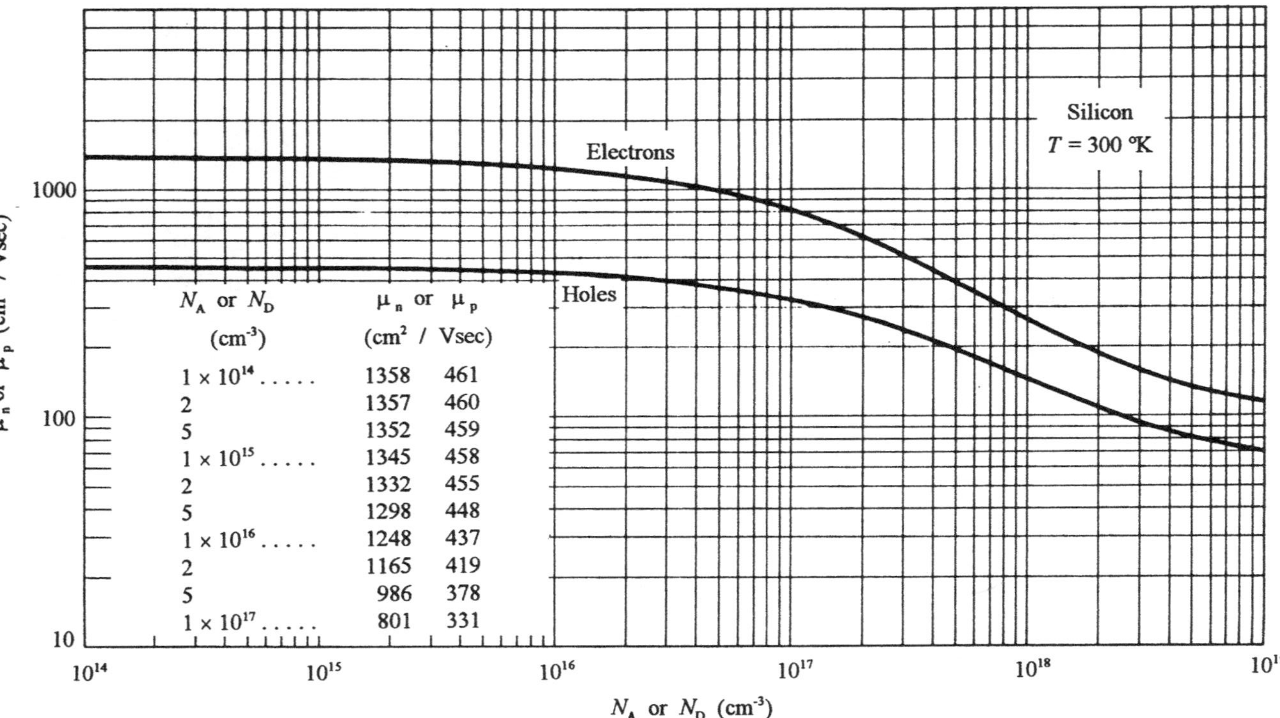

N_A or N_D (cm^{-3})	μ_n or μ_p (cm^2 / Vsec)	
1×10^{14}	1358	461
2	1357	460
5	1352	459
1×10^{15}	1345	458
2	1332	455
5	1298	448
1×10^{16}	1248	437
2	1165	419
5	986	378
1×10^{17}	801	331

Figure 1.6 The variation of electron and hole mobilities with the net doping concentration in silicon. (From Pierret [2]. Copyright 1987 by Addison Wesley Publishing Company, Inc. Reprinted with permission of the publisher.)

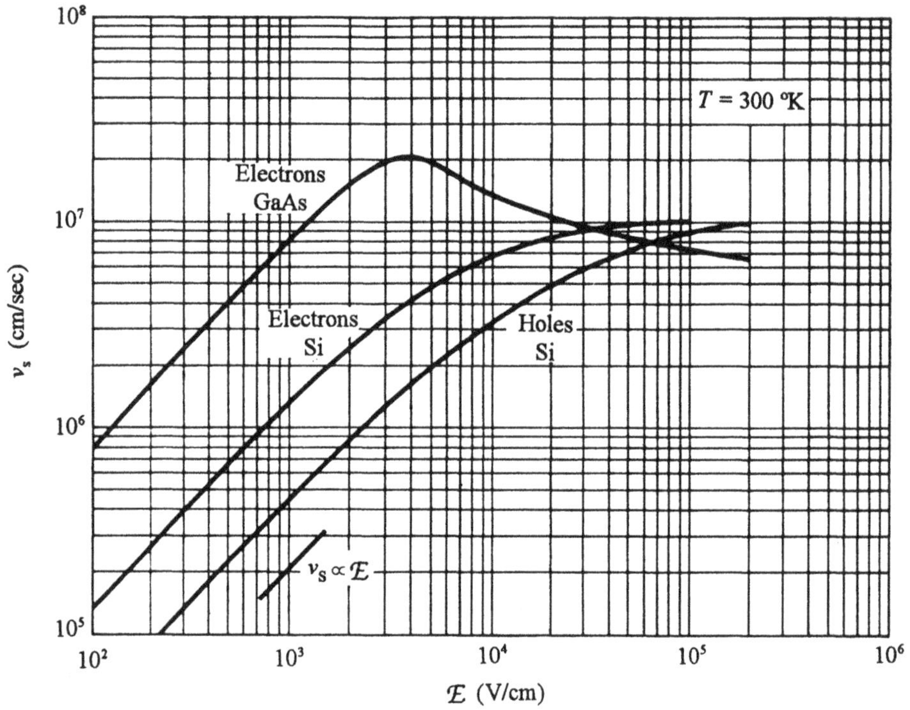

Figure 1.7 Drift velocity as a function of the electric field. (From Pierret [2]. Copyright 1987 by Addison Wesley Publishing Company, Inc. Reprinted with permission of the publisher.)

migration and emigration rates. But, of course, the population of a town can change with time due to unequal rates of birth and death. There is an analogous effect in semiconductors. Free electrons and holes enjoy birth and suffer death through the so-called *generation* and *recombination* events.

In summary, we expect $p(n)$ to vary temporally if $J_p(J_n)$ varies spatially and/or if the rates of the generation and recombination processes are not balanced. The relationship between dp/dt (dn/dt) and these two factors is described by two continuity equations, one applicable to holes and the other to free electrons. Soon we will construct these equations. To be able to do that, however, we first need to familiarize ourselves with the generation and recombination processes in some detail.

Generation and Recombination Processes

We know that a valence electron vacating its bond leaves behind a hole. Where can this electron go? One possibility is to fill in another vacant valence bond, and thus transport a hole in the opposite direction. This, as we know from Figure 1.2, is how the hole current is generated. But, of course, no hole or free electron is created or lost in the course of this process because the electron never ceases to be a valence electron. In terms of the energy-band diagram, the electron leaves one energy level in the valence band to occupy an empty one in the same band. Neither the number of occupied levels in the conduction band (free electrons) nor that of the unoccupied ones in the valence band (holes) is affected.

Now consider the case of a valence electron gaining sufficient energy to exit the valence band. If the absorbed energy exceeds the bandgap energy E_g, the electron will rise to the conduction band and become free. A lesser amount of energy, on the other hand, will take the electron to a trap level inside the bandgap. These two possibilities are illustrated on an energy-band diagram in Figure 1.8(a,b), respectively. The former process, involving direct transition of an electron from the valence band to the conduction band, obviously generates a free-electron/hole pair, and is called *direct generation.* The process depicted in Figure 1.8(b), which involves the elevation of a valence-electron to a trap level, generates a hole but not a free electron, and is called *hole emission.* A third possible generation process, illustrated in Figure 1.8(c), involves the transition of a trapped electron to the conduction band. As in the other two generation processes, the electron must absorb sufficient energy to effect the transition. Because the end product is a free electron, this process is referred to as *electron emission.*

An electron cannot occupy a high energy level forever. Eventually, it releases some of its energy and makes a downward transition toward the valence band. The processes involved are the inverse of those involved in generation, and are commonly called *recombination.*

As shown in Figure 1.8(d), a free electron can lose energy to be trapped and immobilized by the inverse of what we defined above as electron emission. This

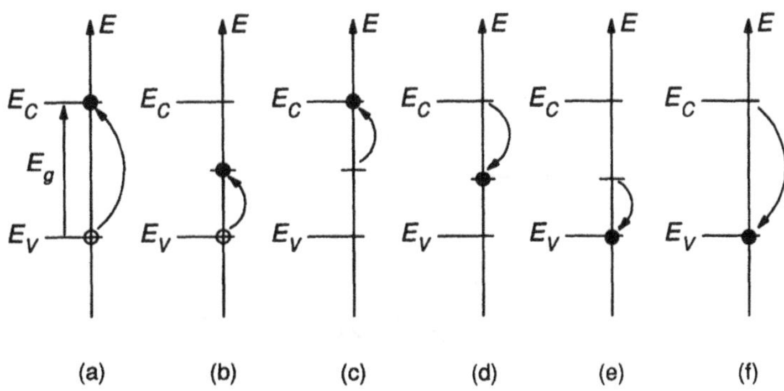

Figure 1.8 Electron transition between the conduction band, valence band, and bandgap levels: (a) direct generation, (b) hole emission, (c) electron emission, (d) electron capture, (e) hole capture, (f) direct recombination.

process, which has the specific name *electron capture,* reduces the free-electron population. It is also possible for a trapped electron to release energy and make the transition to the valence band. As shown in Figure 1.8(e), a hole is removed by this process, which has the specific name *hole capture.* Finally, an electron can make a direct transition from a conduction band level to a valence band level, as depicted in Figure 1.8(f). This process, called *direct recombination,* results in a loss of one free-electron/hole pair.

The two-way direct transition between the conduction and valence bands is the dominant generation and recombination process in compound semiconductors such as GaAs. In silicon, the trap-assisted transition processes dominate. Also note that the energy required for the generation processes is supplied by the thermal reservoir of the crystal or by other sources, such as radiation or particle impact. Recombination events, on the other hand, release energy either back to the thermal reservoir as heat or to the outside of the crystal in the form of photon energy.

Derivation of the Continuity Equations

Consider the semiconductor volume $A\ dx$ shown in Figure 1.9. Suppose (1) the hole current I_p varies by dI_p along dx, (2) the hole recombination rate per unit volume per unit time is R [$\text{cm}^{-3} \cdot \text{s}^{-1}$], and (3) the hole generation rate per unit volume per unit time is $G_{th} + g$, where G_{th} [$\text{cm}^{-3} \cdot \text{s}^{-1}$] and g [$\text{cm}^{-3} \cdot \text{s}^{-1}$] are the rates of the thermal and nonthermal generation processes, respectively. Obviously, the current I_p at x brings into the volume I_p/q holes per unit time, while the current at $x + dx$ removes $(I_p + dI_p)/q$ holes. In the meantime, $(G_{th} + g)A\ dx$ holes are generated, and $RA\ dx$ holes recombine inside the same volume. Therefore, the net increase in hole population per unit time, $d(pA\ dx)/dt$, can be written as

$$\frac{d}{dt}(pA\ dx) = \frac{I_p}{q} - \frac{I_p + dI_p}{q} + (G_{th} + g)A\ dx - RA\ dx$$

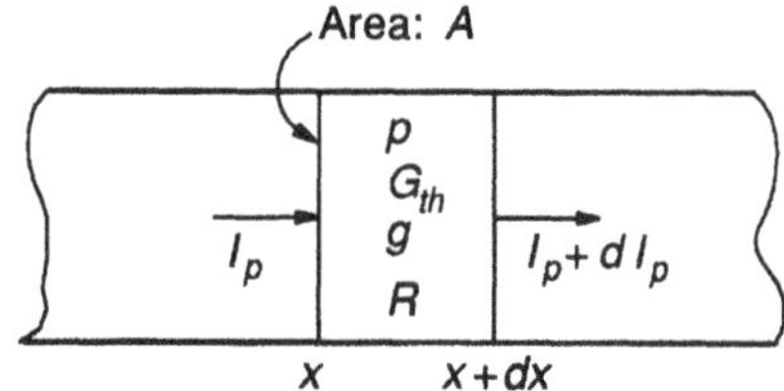

Figure 1.9 A difference dJ_p between the two boundary values of the hole current density and a difference between the generation $(G_{th} + g)$ and recombination (R) rates generally result in a temporal variation in the hole concentration (p) inside the infinitesimal volume Adx. The relationship between these variations is described by the hole continuity equation.

Dividing both sides by the time-independent $A\ dx$, and substituting J_p for I_p/A, we obtain

$$\frac{\partial p}{\partial t} = -\frac{1}{q}\frac{\partial J_p}{\partial x} + (G_{th} + g - R)_h \tag{1.23}$$

where $(G_{th} + g - R)_h$ represents the net generation rate of holes. The foregoing derivation, if repeated for free electrons, will yield

$$\frac{\partial n}{\partial t} = \frac{1}{q}\frac{\partial J_n}{\partial x} + (G_{th} + g - R)_e \tag{1.24}$$

where $(G_{th} + g - R)_e$ is the net generation rate of free electrons. Equations (1.23) and (1.24) are the *continuity equations* for holes and free electrons, respectively. Their multidimensional forms are as follows:

$$\frac{\partial p}{\partial t} = -\frac{1}{q}\nabla \mathbf{J}_p + (G_{th} + g - R)_h$$

$$\frac{\partial n}{\partial t} = \frac{1}{q}\nabla \mathbf{J}_n + (G_{th} + g - R)_e$$

Most problems in device analysis and modeling involve a dc steady state, which implies vanishing time derivatives and, therefore, reduces the continuity equations to

$$\frac{dJ_p}{dx} = q(G_{th} + g - R) \tag{1.25}$$

$$\frac{dJ_n}{dx} = -q(G_{th} + g - R) \tag{1.26}$$

Note that free electrons and holes *must* have the same net generation rate in a dc steady state. To understand why, consider the fact that $(G_{th} + g - R)_h$ represents the net amount of electrons exiting the valence band per unit volume per unit time, whereas $(G_{th} + g - R)_e$ is the net amount of electrons entering the conduction band in the same volume and time. If these two rates were different, the electron population trapped in the bandgap between these two bands would inevitably change with time in an obvious contradiction to what a dc steady state implies. This is why the subscripts h and e are absent in (1.25) and (1.26).

We close this section by pointing out another extremely important implication of the dc steady state. Notice from (1.25) and (1.26) that

$$\frac{dJ_p}{dx} + \frac{dJ_n}{dx} = 0$$

Since $J_p + J_n$ equals the total current density J [see (1.22)], we reach the conclusion

$$\frac{dJ}{dx} = 0 \tag{1.27}$$

that is, total current density in a semiconductor in a dc steady state must be spatially constant.

1.1.5 More on Energy-Band Diagrams

The concept of energy bands was introduced in Section 1.1.1 and then used in Section 1.1.4 as an aid in the description of generation and recombination events. Energy-band diagrams are indeed a very useful tool in device analysis; we need to know more about them.

Figure 1.10 shows an energy-band diagram depicting not only the familiar energy levels E_V and E_C but also the *potential energy level* U and the so-called *intrinsic Fermi energy* E_{Fi}. The latter is located virtually midway between E_V and E_C; its significance will become clear in Section 1.2.1. On the other hand, U can be regarded as the level of energy an electron must attain in order to escape the semiconductor into vacuum and thus disassociate itself from the semiconductor. The difference $\kappa \equiv U - E_C$, which is called *electron affinity,* is a parameter determined by the properties of the semiconductor material.

The potential energy U is related to the electrostatic potential ψ by

$$U = -q\psi \tag{1.28}$$

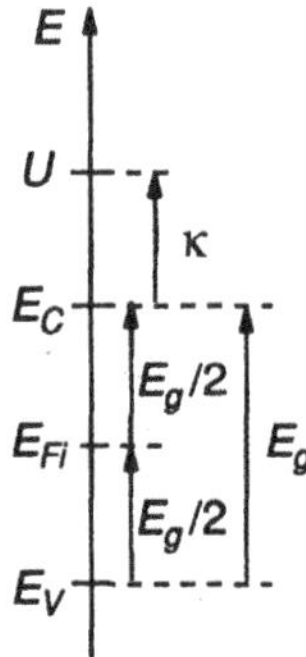

Figure 1.10 An energy-band diagram depicting the potential energy level U, the intrinsic Fermi energy E_{Fi}, and the electron affinity κ.

Note that the units of energy, charge, and potential are the joule, coulomb, and volt, respectively. Using the universal value of q in (1.28), we obtain

$$U\,[\mathrm{J}] = -1.6 \times 10^{-19}\,[\mathrm{C}] \times \psi\,[\mathrm{V}]$$

More often than not, we prefer to use *electron-volt* as a unit of energy. Since $U\,[\mathrm{J}] = 1.6 \times 10^{-19}\,[\mathrm{J/eV}] \times U\,[\mathrm{eV}]$, the above equation can be rewritten as

$$1.6 \times 10^{-19}\,[\mathrm{J/eV}] \times U\,[\mathrm{eV}] = -1.6 \times 10^{-19}\,[\mathrm{C}] \times \psi\,[\mathrm{V}]$$

which can be rearranged into

$$U\,[\mathrm{eV}] = -1\left[\frac{\mathrm{C}}{\mathrm{J}}\mathrm{eV}\right] \times \psi\,[\mathrm{V}]$$

But $[\mathrm{C/J}] = [1/\mathrm{V}]$. Therefore,

$$U\,[\mathrm{eV}] = -1\left[\frac{\mathrm{eV}}{\mathrm{V}}\right] \times \psi\,[\mathrm{V}]$$

according to which, the potential energy in units of electron-volts is just -1 times the electrostatic potential in units of volts. For example, $\psi = 2\mathrm{V}$ corresponds to $U = -2$ eV, and $\psi = -3.2\mathrm{V}$ corresponds to $U = 3.2$ eV.

According to (1.6), a nonzero electric field implies a position-dependent electrostatic potential, which, in turn, implies through (1.28) a position-dependent potential energy; that is

$$\mathscr{E} = -\frac{d\psi}{dx} = \frac{1}{q}\frac{dU}{dx} \tag{1.29}$$

The energy-band diagram of a semiconductor in which $\mathscr{E} \neq 0$ is therefore generally position dependent as well. For this reason, we usually draw energy-band diagrams on a two-dimensional E-x coordinate system, as illustrated in Figure 1.11(a) for an arbitrary region in which E_g and κ are also assumed to be x dependent. Such diagrams convey a wealth of information about the semiconductor region. For example, the magnitude and direction of $\mathscr{E}$ at a given point can be immediately determined from the slope of the $U(x)$ curve, as implied by (1.29). A negative slope, such as that at point x_1 in Figure 1.11(a), corresponds to a field directed opposite to x. A positive slope, on the other hand, indicates a field in the same direction as x; point x_2 in Figure 1.11(a) is an example.

Energy-band diagrams also allow us to determine electrostatic-potential difference between any two points of the semiconductor, say, point x_3 and point x_4

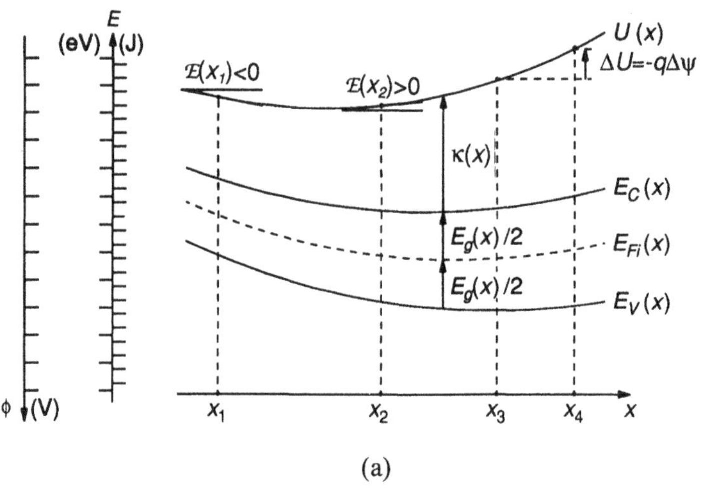

(a)

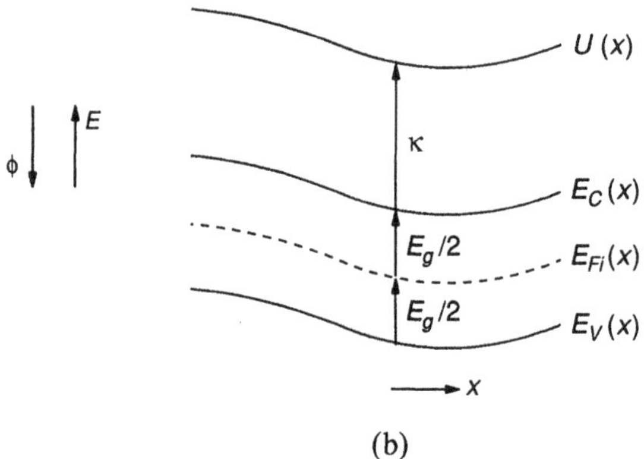

(b)

Figure 1.11 (a) A two-dimensional energy-band diagram. Notice the dual calibration of the energy axis and the possibility of interpreting the levels in terms of potential (ϕ) instead of energy (E). (b) In a homogeneous semiconductor, in which both E_g and κ are spatially constant, the bands remain equidistant in energy.

of Figure 1.11(a). All we need to do is to divide by $-q$ the potential-energy difference between these two points, that is

$$\psi(x_4) - \psi(x_3) = -\frac{1}{q}[U(x_4) - U(x_3)]$$

To be able to read electrostatic-potential differences directly, we sometimes add a second vertical axis to energy-band diagrams to represent the general variable of potential ϕ. We create this axis by reversing the direction of the E axis, and scaling it by q in accordance with the transformation

$$\Delta\phi = -\frac{\Delta E}{q} \tag{1.30}$$

The difference between the ϕ coordinates of any two points on the $U(x)$ curve now gives us directly the electrostatic-potential difference between these two points. Furthermore, we can now directly interpret the difference between any two energy levels at a given point of a semiconductor as a potential difference. For example, the bandgap, which is originally defined as the difference between E_C and E_V, and is measured in units of electron-volts or joules, can be expressed in terms of the potential difference between these two energy levels and be measured in units of volts. How we benefit from this transformation will become abundantly clear in the following section.

The semiconductor regions in which E_g or κ are position dependent are called *heterogeneous* regions. Heterogeneity may be introduced intentionally, as in the emitter-base junction of a heterojunction bipolar transistor (HBT), or unintentionally, as in the highly doped emitter of a conventional BJT. Except for the BJT emitter, however, the regions that make up the conventional devices can be regarded as *homogeneous* regions, in which E_g and κ are independent of x. As illustrated in Figure 1.11(b), the position dependence of E_C, E_V, and E_{Fi} in a homogeneous region must match the position dependence of U; that is

$$\mathscr{E} = -\frac{d\psi}{dx} = \frac{1}{q}\frac{dU}{dx} = \frac{1}{q}\frac{dE_C}{dx} = \frac{1}{q}\frac{dE_V}{dx} = \frac{1}{q}\frac{dE_{Fi}}{dx} \tag{1.31}$$

Therefore, one can determine the direction of $\mathscr{E}$, and therefore the gradient of ψ, in a homogeneous region by inspecting the position dependence of E_C, E_V, or E_{Fi}. As a matter of fact, we usually omit U in the energy-band diagram of such a region, and still convey full information on bands using the plots of E_C, E_V, and E_{Fi}.

1.2 THERMAL EQUILIBRIUM

As long as no net energy transfer occurs between a semiconductor and its environment, the semiconductor remains in a *thermal equilibrium state.* In other words, thermal equilibrium is the unactivated state of a semiconductor. A device that passes current, or is heated to a temperature different than the ambient, or receiving radiation or energetic particles, obviously, does not fit into this definition of the thermal equilibrium state; but a device standing in an unpowered circuit usually does. Some of us will now wonder why in device analysis and modeling we should be concerned with the thermal equilibrium state, in which a device is inactive by definition. It is true that our main objective is to analyze and to model device behavior in response to external excitations, which usually force the device into a *nonequilibrium state.* However, treating the case of nonequilibrium as a departure from equilibrium greatly facilitates the analysis. This is why the state of equilibrium is not as trivial as it first appears to be.

In this section, we will see how the five fundamental equations and other general tools presented in Section 1.1 can be reduced into rather primitive and, hence, manageable forms for the case of thermal equilibrium. The conclusions will then be applied to a general analysis of semiconductor regions in equilibrium.

1.2.1 General Equilibrium Properties of Semiconductors

Fundamental Equations in Equilibrium

The state of thermal equilibrium implies a steady state, hence

$$\frac{dp}{dt} = \frac{dn}{dt} = 0 \tag{1.32}$$

as well as an absence of external energy sources, which otherwise could activate nonthermal generation events, hence

$$g = 0 \tag{1.33}$$

It also implies an absence of carrier transport, which otherwise would lead to energy consumption, hence

$$J_p = J_n = 0 \tag{1.34}$$

Under these conditions, the total energy of a semiconductor remains constant. Carriers are thermally generated at a rate G_{th} using this invariable reservoir of energy, and in the meantime, recombine at a rate R_o, and thus return to the reservoir the energy they borrowed for generation. To some of us, the equality of G_{th} and R_o may be intuitively obvious. For others, it suffices to substitute (1.33) and (1.34) into any of the two steady-state continuity equations (1.25) and (1.26), to obtain

$$G_{th} = R_o \tag{1.35}$$

As we show later, the value of G_{th} does not depend on whether the semiconductor is in an equilibrium state or in a nonequilibrium one. The recombination rate, on the other hand, depends on the prevailing state. For this reason, we designate its equilibrium value, and for that matter, the equilibrium value of most internal variables, with a subscript o.

Equation (1.35) is the equilibrium form of the two continuity equations. What about the other three fundamental equations? None of the conditions of (1.32), (1.33) and (1.34) affects the form of Poisson's equation, hence

$$\frac{d\mathscr{E}}{dx} = \frac{q}{\epsilon}(p_o - n_o + N) \tag{1.36}$$

As to the equilibrium forms of the current density equations, we can write from (1.13), (1.18), and (1.34) the following:

$$\mathscr{E} = -\frac{kT}{q}\frac{dn_o}{n_o dx} \tag{1.37}$$

$$\mathscr{E} = \frac{kT}{q}\frac{dp_o}{p_o dx} \tag{1.38}$$

The Law of Mass Action

Canceling $\mathscr{E}$ between (1.37) and (1.38) and rearranging, we obtain

$$\frac{1}{p_o n_o}\frac{d(p_o n_o)}{dx} = 0$$

which implies that the product $n_o p_o$ is position independent. A detailed quantum-mechanical derivation also taking (1.35) into account yields

$$p_o n_o = n_i^2 \tag{1.39}$$

where the position-independent quantity

$$n_i \equiv CT^{3/2} \exp(-E_g/2kT) \tag{1.40}$$

is called *intrinsic carrier concentration.* Note that C is a constant and E_g is the bandgap energy. The value of n_i is given in Figure 1.12 as a function of temperature for three semiconductors of different bandgap energies.

The relationship between the equilibrium concentrations of electrons and holes, as defined by (1.39), is called the *law of mass action.* It has tremendous utility in the equilibrium analysis of semiconductor devices.

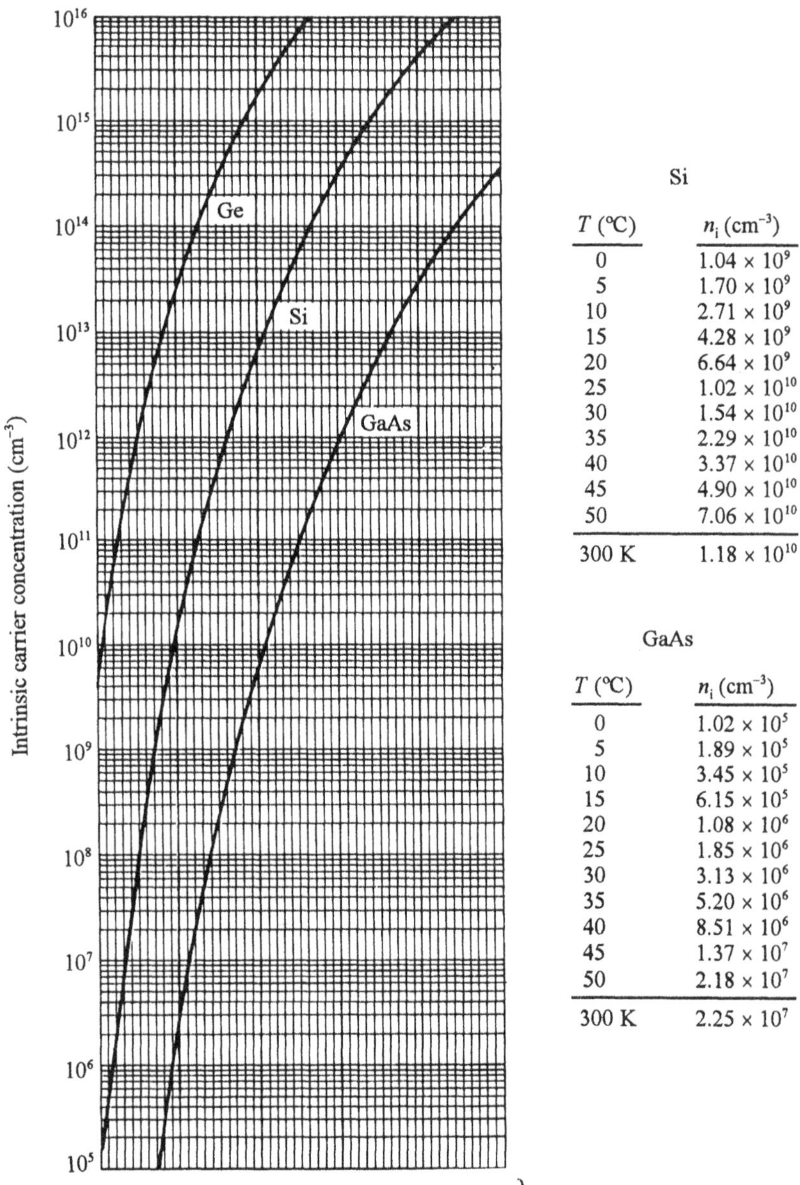

Si

T (°C)	n_i (cm^{-3})
0	1.04×10^{9}
5	1.70×10^{9}
10	2.71×10^{9}
15	4.28×10^{9}
20	6.64×10^{9}
25	1.02×10^{10}
30	1.54×10^{10}
35	2.29×10^{10}
40	3.37×10^{10}
45	4.90×10^{10}
50	7.06×10^{10}
300 K	1.18×10^{10}

GaAs

T (°C)	n_i (cm^{-3})
0	1.02×10^{5}
5	1.89×10^{5}
10	3.45×10^{5}
15	6.15×10^{5}
20	1.08×10^{6}
25	1.85×10^{6}
30	3.13×10^{6}
35	5.20×10^{6}
40	8.51×10^{6}
45	1.37×10^{7}
50	2.18×10^{7}
300 K	2.25×10^{7}

Figure 1.12 The temperature dependence of the intrinsic carrier concentration in germanium, silicon, and gallium arsenide. (From Pierret [1]. Copyright 1988 by Addison Wesley Publishing Company, Inc. Reprinted with permission of the publisher.)

The Fermi Formalism

The above-mentioned quantum-mechanical derivation also introduces a formalism by which the carrier concentrations are represented as follows:

$$n_o = n_i \exp\left(\frac{E_F - E_{Fi}}{kT}\right) \tag{1.41}$$

$$p_o = n_i \exp\left(-\frac{E_F - E_{Fi}}{kT}\right) \tag{1.42}$$

where E_F is the *Fermi energy,* and E_{Fi} is the intrinsic Fermi energy, which was introduced in Section 1.1.5 as a level located virtually midway between E_C and E_V.

It is important to understand that (1.41) and (1.42) are independent of Poisson's equation but not of the other fundamental equations (1.35), (1.37), and (1.38). Equations (1.41) and (1.42) can be regarded as a mathematically legitimate decomposition of the law of mass action relationship (1.39). The latter implies that a single variable suffices to represent both n_o and p_o, which is what E_F does in (1.41) and (1.42).

According to (1.41) and (1.42), an E_F lying above E_{Fi} on an energy-band diagram indicates $n_o > p_o$. At any region of the semiconductor, where this condition is satisfied, electrons are the *majority carrier* and holes are the *minority carrier.* Such a region is said to have *n-type conductivity.* Obviously, the more positive the energy difference $E_F - E_{Fi}$, the larger the concentration of electrons. The opposite condition, $E_F - E_{Fi} < 0$, indicates $p_o > n_o$, hence a majority of holes, and thus *p-type conductivity.* An increasingly negative $E_F - E_{Fi}$ corresponds to an increasing hole concentration. A third possibility is $E_F = E_{Fi}$, which implies $p_o = n_o$. These cases are shown in Figures 1.13(a–c). Notice the representation of the energy difference $E_F - E_{Fi}$ with an arrow marked ϕ_F. The latter is called *Fermi potential*; it is the potential difference related to $E_F - E_{Fi}$ by (1.30), that is

$$\phi_F \equiv -\frac{E_F - E_{Fi}}{q} \tag{1.43}$$

which transforms (1.41) and (1.42) into

$$n_o = n_i \exp[-(q/kT)\phi_F] \tag{1.44}$$

$$p_o = n_i \exp[(q/kT)\phi_F] \tag{1.45}$$

Expressing n_o and p_o in terms of a potential difference instead of an energy difference is a convenience because, in device modeling, one eventually needs to relate internal variables to port variables, which usually include potentials but not energies. Also note that an upward arrow, like the ϕ_F of the energy-band diagram of Figure 1.13(a), indicates a negative potential difference but a positive energy difference. The arrow of ϕ_F in Figure 1.13(b) indicates just the opposite. Of course, we remember from Section 1.1.5 that ϕ and E have opposite reference directions.

As long as E_F lies within the bandgap, that is, $|E_F - E_{Fi}| < E_g/2$, the semiconductor is said to be *nondegenerate*. In the limit of nondegeneracy, $|E_F - E_{Fi}| = E_g/2$, the majority-carrier concentration reaches a very large value. For a larger majority-carrier concentration, E_F enters either the conduction band or the valence band. Then, the semiconductor becomes *degenerate,* which is a condition necessitating significant modifications in our mathematical arsenal. Unless otherwise stated, we will assume a nondegenerate semiconductor in our entire study.

The Fermi formalism, introduced with (1.41) and (1.42) and discussed thereafter, may appear to be an unnecessary complication because, after all, it does not offer any relationship independent of the law of mass action or of the fundamental equations. Indeed, it does not offer anything new in that respect, but it provides us with a vehicle, E_F, by which the carrier concentrations n_o and p_o, can be represented graphically on energy-band diagrams. We need to be patient to see how greatly this formalism can facilitate the equilibrium analysis of semiconductor devices. As a matter of fact, later we will extend it to nonequilibrium analyses as well.

A few more general conclusions about E_F and ϕ_F are in order. First, substituting (1.41) together with $\mathscr{E} = (1/q)dE_{Fi}/dx$ of (1.31) into (1.37), we arrive at the extremely important conclusion

$$\frac{dE_F}{dx} = 0 \tag{1.46}$$

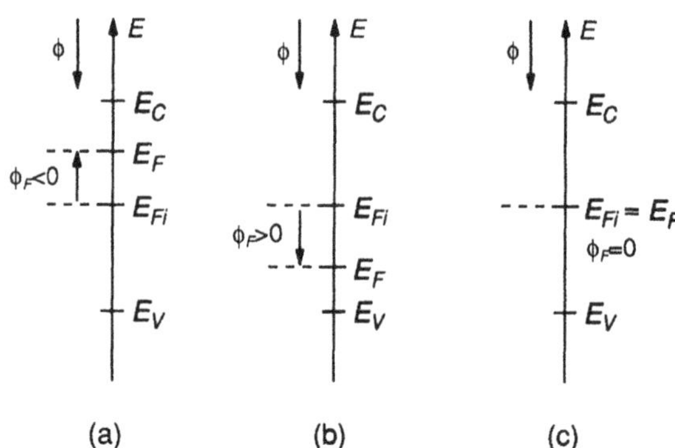

Figure 1.13 Energy-band diagrams depicting (a) an *n*-type point, (b) a *p*-type point, and (c) an intrinsic point.

that is, the Fermi energy is independent of position.[2] Using this in (1.43), we obtain

$$\frac{d\phi_F}{dx} = \frac{1}{q}\frac{dE_{Fi}}{dx}$$

which implies from (1.31) the following relationships:

$$\mathscr{E} = \frac{d\phi_F}{dx} \tag{1.47}$$

and

$$\phi_F + \psi = \text{constant} \tag{1.48}$$

The constancy of E_F and $\phi_F + \psi$ is illustrated in Figure 1.14 on the energy-band diagram of an arbitrary (but homogeneous) semiconductor region in equilibrium. Also notice the parallel variation of E_C, E_V, and E_{Fi} with position.

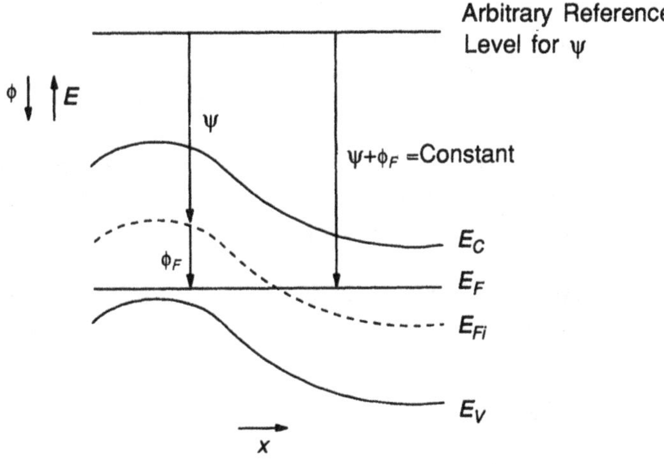

Figure 1.14 The general features of the equilibrium energy-band diagram of a homogeneous semiconductor region.

[2]Repeating the calculation with (1.42) and (1.38) would yield the same result. Also note that the constancy of E_F is a direct consequence of the second law of thermodynamics; it is valid for any system in thermal equilibrium.

1.2.2 Analysis in Equilibrium

The General Equation of the Equilibrium State

In an equilibrium analysis problem, we are usually provided with the structural parameters of the device involved, such as the doping-concentration profiles and geometric features, and are asked to compute one or more of the internal variables, such as n_o, p_o, ϕ_F, ψ, $\mathscr{E}$, or ρ, as functions of position or of temperature for a given set of boundary conditions.

Note that the equilibrium state of any semiconductor device can be described by a single differential equation in terms of any of the previously mentioned internal variables. For example, rewriting Poisson's equation (1.36) by substituting (1.47) for $\mathscr{E}$, (1.45) for p_o, and (1.44) for n_o, we obtain the general equation of equilibrium in terms of ϕ_F as follows:

$$\frac{d^2\phi_F}{dx^2} = \frac{q}{\epsilon}\left[n_i \exp\left(\frac{q}{kT}\phi_F\right) - n_i \exp\left(-\frac{q}{kT}\phi_F\right) + N\right] \tag{1.49}$$

Solving this equation with the appropriate doping profile and boundary conditions would yield the Fermi potential profile $\phi_F(x)$. Then, one could obtain $n_o(x)$ from (1.44), $p_o(x)$ from (1.45), ψ from (1.48), $\mathscr{E}$ from (1.47), and ρ from (1.4). However, the general equation of equilibrium, regardless of the internal variable it represents, can yield an analytical solution only for *neutral regions* in which $\rho = 0$. All conventional devices, including BJTs and MOSFETs, do indeed contain neutral regions, but they also contain *space-charge regions* in which $\rho \neq 0$. Although the general equation does not yield an exact solution for space-charge regions, approximate solutions based on certain simplifying assumptions are widely used. We will present the solutions as needed in later chapters.

In the remainder of this section, the solutions for neutral regions will first be presented. Subsequently, the formation of neutral regions and space-charge regions will be discussed with a view toward identifying their location in device structures.

Solutions for Neutral Regions

The definition $\rho = 0$ of neutrality reduces (1.4) to

$$p_o - n_o + N = 0 \tag{1.50}$$

Rewriting this equation with $p_o = n_i^2/n_o$ of the law of mass action (1.39), and then, solving for n_o, we obtain

$$n_o = \frac{1}{2}N + \sqrt{\frac{1}{4}N^2 + n_i^2} \qquad (1.51)$$

for the electron concentration in a neutral region in equilibrium.

We will soon determine that, in a neutral region, $|N|$ should be several orders of magnitude greater than n_i. This fact enables us to simplify (1.51) into

$$n_o = N \qquad \text{for } N > 0 \qquad (1.52)$$

$$n_o = \frac{n_i^2}{|N|} \qquad \text{for } N < 0 \qquad (1.53)$$

The corresponding hole concentrations, as derived from these equations and (1.39) are as follows:

$$p_o = \frac{n_i^2}{N} \qquad \text{for } N > 0 \qquad (1.54)$$

$$p_o = |N| \qquad \text{for } N < 0 \qquad (1.55)$$

Once n_o and p_o have been determined from these equations, one can obtain ϕ_F from (1.44) or (1.45). The remaining internal variables are then derived from ϕ_F.

Equations (1.52), (1.53), (1.54), and (1.55) lead us to the following conclusions regarding neutral regions in equilibrium:

1. Electrons are the majority carriers if $N > 0$.
2. Holes are the majority carriers if $N < 0$.
3. The majority-carrier concentration is virtually equal to $|N|$.
4. The minority-carrier concentration $n_i^2/|N|$ is negligibly small in comparison with the majority-carrier concentration.

These very simple results are extremely useful in the equilibrium analysis of devices. But, of course, they are valid only in neutral regions. Therefore, we need to know where in devices the neutral regions can exist. This is what we will discuss next.

Formation of Neutral Regions and Space-Charge Regions

Let us write the general equation of equilibrium in terms of n_o. Starting with Poisson's equation, substituting (1.37) for $\mathscr{E}$, using n_i^2/n_o for p_o, and finally solving for n_o, we obtain

$$n_o = \left(1 + \frac{kT\epsilon}{q^2N}\frac{d^2 \ln n_o}{dx^2}\right)\frac{N}{2} + \sqrt{\left(1 + \frac{kT\epsilon}{q^2N}\frac{d^2 \ln n_o}{dx^2}\right)^2\frac{N^2}{4} + n_i^2} \qquad (1.56)$$

This equation, being absolutely general, is valid everywhere in a semiconductor region in equilibrium. Comparing it with (1.51) of neutral regions, we immediately realize that a region in equilibrium can be regarded as neutral if

$$L_D^2 \left| \frac{d^2 \ln n_o}{dx^2} \right| \ll 1 \tag{1.57}$$

where

$$L_D \equiv \sqrt{\frac{kT\epsilon}{q^2|N|}} \tag{1.58}$$

is the so-called *Debye length.* But we also know that, if a region is neutral, n_o, will be related to N through (1.51). Using the latter for n_o in (1.57), we find

$$L_D^2 \left| \frac{d^2}{dx^2} \ln\left(\frac{1}{2}N + \sqrt{\frac{1}{4}N^2 + n_i^2}\right) \right| \ll 1 \tag{1.59}$$

as a necessary condition to be satisfied by the doping profile in order for the region to be neutral. For a given $N(x)$, therefore, (1.59) can be used as a test for neutrality. We will now use this test to analyze the neutrality of such common substructures of conventional devices as *pn* junctions, high-low junctions, uniformly doped regions, and regions of nonzero doping gradient. The BJT contains all of these features and, therefore, can serve as a sufficiently general medium for illustrating the results.

Figure 1.15(a) shows a typical *npn*-type BJT doping profile taken along the emitter/buried-layer axis where the main transistor action takes place. Note that N is positive in the emitter and collector/buried-layer regions. The former region is located between $x = 0$ and $x = 0.75$ μm, whereas the latter region extends beyond $x = 1.85$ μm. Between the two, where the base region is located, N is negative. By definition, a *pn* junction is formed where N changes polarity. The BJT structure has two such junctions demarcating the emitter, base, and collector regions. The gradual change in the magnitude of N between $x = 8$ and $x = 11$ μm marks the location of a *high-low* junction. This is where the collector meets the buried layer. In Chapter 2 we will learn more about these regions.

The left-hand side of (1.59), as calculated from the profile of Figure 1.15(a), is plotted in Figure 1.15(b). Knowing that the higher the plotted value, the worse the neutrality, we arrive at the following conclusions:

1. The neighborhoods of *pn* junctions are definitely space charged because the plotted value approaches infinity there.
2. Although parts of the high-low junction come close to being space charged, it is reasonable to assume neutrality there because the plot remains at least two orders of magnitude less than unity.

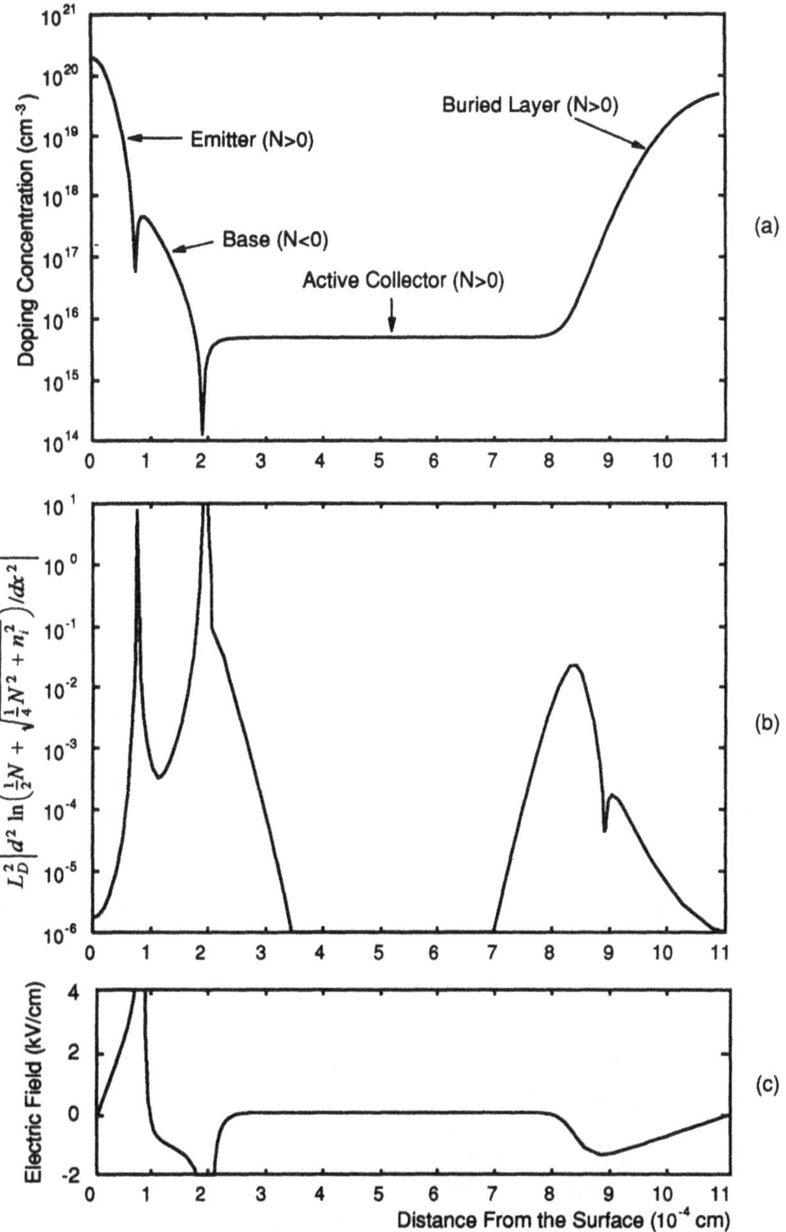

Figure 1.15 The formation of neutral and space-charged regions in a BJT structure: (a) A typical doping concentration profile taken along the emitter/buried-layer axis. (b) The left-hand side of (1.59), as compared to unity, is an indicator of the extent of neutrality. The smaller the plotted value, the more justifiable the assumption of neutrality. (c) The electric-field profile.

3. The emitter and base regions, excluding the *pn* junction neighborhoods, are neutral to a good approximation.
4. The collector region is almost perfectly neutral.
5. In the neutral regions, $|N|$ is many orders of magnitude greater than n_i whose room temperature value for silicon is 1.1×10^{10} cm^{-3}.

Relying on the last of these conclusions, we can express the electric field in a neutral region simply by substituting (1.52) or (1.53) into (1.37). The result is

$$\mathscr{E} = -\frac{kT}{q}\frac{1}{N}\frac{d|N|}{dx} \tag{1.60}$$

Figure 1.15(c) gives the electric field profile as calculated from this equation for the neutral regions of the BJT. Notice that the field vanishes in the "perfectly" neutral collector bulk. This is a result of the uniform doping profile. In the "approximately" neutral emitter, base, and high-low junction regions, the doping profile is nonuniform; therefore, the field is nonzero.

For a physical explanation of the presence of an electric field in nonuniformly doped regions, consider, for example, the emitter where N is a decreasing function of x. Since the region is neutral, the majority-carrier concentration profile $n_o(x)$ is virtually identical to $N(x)$, hence, it is a decreasing function of x. On the other hand, the minority-carrier concentration $p_o(x)$, being virtually identical to $n_i^2/N(x)$, is an increasing function of x. Due to these carrier concentration gradients, electrons tend to diffuse to the right, and thus conduct a diffusion current to the left, whereas holes tend to diffuse to the left, and thus conduct a diffusion current in the same direction. In equilibrium, however, neither type of carriers can conduct current, which implies the presence of drift currents to compensate for the diffusion currents. Drift, in turn, necessitates an electric field, which must be directed to the right, that is, $\mathscr{E} > 0$, in order to yield drift currents opposing the previously mentioned diffusion currents. As observed in Figure 1.15(c), the field is indeed positive in the emitter region. Let us now pose the question as to how this field develops. It develops simply by n_o being slightly lower than N, and p_o being slightly higher than n_i^2/N. This makes the volumetric charge density $\rho \equiv (p_o - n_o + N)$ positive, and thus causes $\mathscr{E}$ to increase with position as dictated by Poisson's equation. However, the difference between N and n_o is so small in comparison with N itself that n_o is still approximately equal to N, which also implies $p_o \cong n_i^2/N$. This is why the assumption of neutrality in the emitter region is only an approximation, albeit a good one.

The reader is encouraged to examine from the same physical perspective the neutrality of the base, collector, and high-low junction regions as well.

According to (1.60), the electric field in a neutral region in equilibrium is determined solely by the doping profile and temperature. This must be true for the boundaries of the region as well. But, on the other hand, the boundary value of the electric field can be imposed externally, and therefore may not comply with what is dictated by (1.60). Does this contradiction preclude neutrality at the neighborhood of a boundary? Yes, it generally does! If the neutrality is to extend all the way to the boundary, not only must the doping profile satisfy (1.59), but the externally imposed boundary value of the field, say, $\mathscr{E}(0)$, must also satisfy the condition

$$\mathscr{E}(0) = \mathscr{E}_N(0) \tag{1.61}$$

where $\mathscr{E}_N$ is the field dictated by (1.60). Unless (1.61) is satisfied, a space-charge region develops at the boundary neighborhood. As shown in Figure 1.16, the difference $\mathscr{E}(x) - \mathscr{E}_N(x)$ decreases with position along this space-charge region and eventually vanishes at some depth where neutrality is restored. The formation and manipulation of such a *surface space-charge region* is the essence of operation in all MOS structures; we will examine this formation in detail in Chapter 3.

We will now close this section by restating the main conclusions regarding neutrality in equilibrium:

1. The neighborhood of a *pn* junction is space charged.
2. Surface neighborhoods are generally space charged; but they become neutral if the boundary value of the electric field satisfies (1.61).

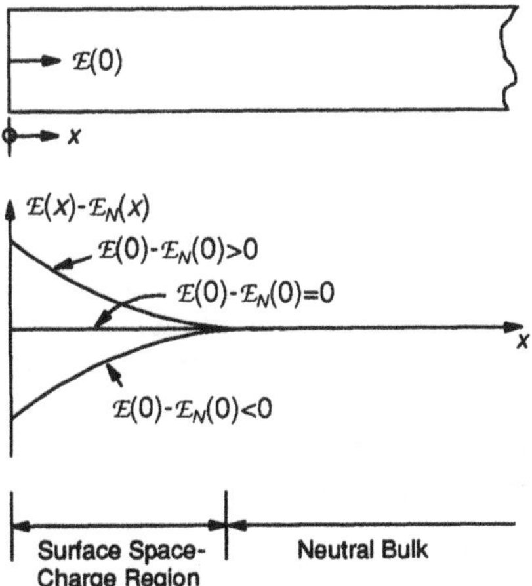

Figure 1.16 The formation of a surface space-charge region when the externally imposed boundary field $\mathscr{E}(0)$ differs from the value dictated by (1.60). Notice that the difference decays with position, leading to the restoration of neutrality inside the semiconductor.

3. Nonuniformly doped regions are approximately neutral. The electric field in such regions is nonzero but small.
4. Uniformly doped regions are perfectly neutral and field free.

In the rest of the text, we will use the term *bulk* to designate a region that is neutral in thermal equilibrium.

1.3 NONEQUILIBRIUM

Any physical event that transfers energy between a semiconductor and its environment upsets the thermal equilibrium state of the semiconductor. For this reason, a device operating as a part of an activated circuit is generally in a nonequilibrium state, in which the relatively simple equilibrium forms of the fundamental equations as well as the law of mass action and the related concept of Fermi level are no more valid. Nevertheless, we can still manage to analyze bulk regions with reasonable simplicity provided that the departure from equilibrium is not too large. This, fortunately, suffices for developing basic but useful models for transistors in nonequilibrium. The cases in which the departure from equilibrium is large are then treated by properly augmenting these basic models.

In this section, we will introduce the concept of *injection level* as a measure of departure from equilibrium in a bulk region, and will develop a mathematical basis for an analytical treatment of such a region operating under *low-level injection* conditions in steady state. In addition, the concept of generation and recombination will be revisited with a quantitative model of trapping and, moreover, the Fermi formalism will be extended to nonequilibrium.

1.3.1 Injection Level

A typical consequence of nonequilibrium is the deviation of the two carrier concentrations from their respective equilibrium levels. Carriers are added to or extracted from the semiconductor under the influence of energy-transferring events. The resulting perturbation of carrier concentrations in a bulk region is known as *injection,* and the difference between the nonequilibrium and equilibrium concentrations is called *excess concentration.* We generally use the symbols p' and n' to denote excess concentrations, that is

$$p' \equiv p - p_o \tag{1.62}$$

for the hole excess concentration, and

$$n' \equiv n - n_o \tag{1.63}$$

for the electron excess concentration. In these equations, p and n denote nonequilibrium concentrations. It is important to note that while n_o, p_o, n, and p are always non-negative, p' and n' may be positive or negative, depending on whether the carriers are added to or extracted from the semiconductor.

We can now define the extremely important concept of low-level injection. If, at a given point of a bulk region, the magnitudes of the two types of excess concentrations are much smaller than the doping concentration, then the perturbation at that point is said to be a low-level injection. Mathematically, the condition of low-level injection can be defined as

$$|n'| \ll |N| \qquad \text{and} \qquad |p'| \ll |N| \tag{1.64}$$

Since the majority-carrier concentration in a bulk region in equilibrium equals the doping concentration $|N|$ [see (1.52) and (1.55)], and since the excess concentrations are much smaller than $|N|$ in low-level injection, we expect the majority-carrier concentration to be virtually unaffected by injection if the latter is low; that is,

$$n \cong n_o = N \qquad \text{for } N > 0 \tag{1.65}$$

and

$$p \cong p_o = -N \qquad \text{for } N < 0 \tag{1.66}$$

As for the minority carriers of the equilibrium state, we expect them to remain as minority carriers also in the nonequilibrium state because their equilibrium concentration is negligibly small and their excess concentration is much smaller than $|N|$, as implied by definition (1.64) of the low-level injection condition. In consequence, the type of conductivity (n-type or p-type) of a bulk region does not change with injection if the latter is low.

One more consequence of the low-level injection condition is of prime importance to us; namely, the dependence of the net generation rate on the excess concentration of the minority carriers. To be able to determine the form of this dependence, however, we need to learn the Shockley-Read-Hall theory of carrier trapping first.

1.3.2 The Shockley-Read-Hall Theory of Trapping

In Section 1.1.4, we introduced two net generation rates, $(G_{th} + g - R)_h$ for holes and $(G_{th} + g - R)_e$ for electrons, and subsequently used these rates in constructing the continuity equations. Under normal operating conditions, the nonthermal generation rate g is expected to be negligible, if not zero, everywhere in a transistor.

This, of course, is not the case in photonic transistors, in which photogeneration is the essence of operation. But such devices are not a subject of this text. In conventional transistors, the most notable physical process that involves $g \neq 0$ is the so-called *impact ionization* process, which is discussed in later chapters as a limiting process in BJTs and MOSFETs. Currently, we will assume $g = 0$ and will relate the remainder $G_{th} - R$ of the net generation rate to trapping properties in a quantitative model, which was originally developed by Shockley, Read, and Hall [8,9].

Since direct band-to-band transitions are unlikely to occur in silicon, the generation and recombination of carriers in conventional transistors are dominated by the four trap-assisted electron transition processes as depicted in Figure 1.8(b–e) for a single trap at energy level E. Generally, a large number of such traps are distributed in energy within the bandgap. Consider the traps located within an infinitesimal energy range dE around an arbitrary energy level E. The number of electrons captured from the conduction band by these traps per unit time per unit volume can be represented by $r_{ec}dE$, where r_{ec} is the rate of the electron capture process per unit time per unit volume per unit energy range. The total rate of electron capture (per unit time per unit volume) by all the traps in the entire energy range $E_V \leq E \leq E_C$ of the bandgap is then found by integrating the above rate over the bandgap. Therefore, the total electron capture rate can be expressed as

$$R_{ec} = \int_{E_V}^{E_C} r_{ec}\, dE = \int_{-E_g/2}^{E_g/2} r_{ec}\, d(E - E_{Fi})$$

Changing the variable of integration from E to $E - E_{Fi}$, as in the latter part of this equation, is a mathematical convenience that enables us to take the intrinsic Fermi level as an appropriate reference level to measure the energy of a trap. Of course, the intrinsic Fermi level is located at midband.

The total rates per unit time per unit volume of the remaining three transition processes can be expressed similarly as follows.

Total electron emission rate:

$$R_{ee} = \int_{-E_g/2}^{E_g/2} r_{ee}\, d(E - E_{Fi})$$

Total hole capture rate:

$$R_{hc} = \int_{-E_g/2}^{E_g/2} r_{hc}\, d(E - E_{Fi})$$

Total hole emission rate:

$$R_{he} = \int_{-E_g/2}^{E_g/2} r_{he}\, d(E - E_{Fi})$$

In the Shockley-Read-Hall model, the rates r_{ec}, r_{ee}, r_{hc}, and r_{he} are given by the following equations:

$$r_{ec} = v_{th}\sigma_n n(D_t - n_t) \tag{1.67}$$

$$r_{ee} = v_{th}\sigma_n n_i \exp[(E - E_{Fi})/kT]\, n_t \tag{1.68}$$

$$r_{hc} = v_{th}\sigma_p p n_t \tag{1.69}$$

$$r_{he} = v_{th}\sigma_p n_i \exp[-(E - E_{Fi})/kT]\,(D_t - n_t) \tag{1.70}$$

where D_t is the density (per unit volume per unit energy) of the available traps, n_t is the similarly defined density of the traps occupied by electrons, and σ_n and σ_p are two parameters of trap effectiveness, measured in square centimeters and referred to as *capture cross section* for electrons and holes, respectively. Although we will not examine the derivation of these rate functions, it is instructive to interpret them in terms of the physics of trapping. Note the following in this regard:

1. The total electron capture rate r_{ec} is proportional to the product $n(D_t - n_t)$ because the greater the availability of free electrons and empty traps, the higher the rate of the transition events involving the two. The rate r_{ec} is also proportional to v_{th} because an electron of higher velocity travels a longer distance in unit time, and thus increases its probability of being captured. Finally, the larger the capture cross section of a trap, the greater its ability to capture a passing electron. This is why r_{ec} is proportional to σ_n as well.
2. The total hole capture rate r_{hc} is proportional to the product pn_t because the transition it represents occurs between occupied trap levels of density n_t and empty valence band levels (holes) of density p.
3. The total electron emission rate r_{ee} is associated with a transition from occupied trap levels of density n_t to empty conduction band levels. The latter are so abundant that their density does not limit the process and so does not appear in (1.68). However, the rate is an exponentially increasing function of the trap energy $E - E_{Fi}$ because the closer the trap to the conduction band, the smaller the energy required for effecting the transition.
4. The total hole emission rate r_{he} is determined by $D_t - n_t$ because the hole emission process involves empty traps. Valence electrons, which make the transition in this process, are so plentiful that their density is not a factor in this rate. Also obvious from (1.70) is the strong rate-enhancing effect of the closeness of the trap level to the valence band.

Now consider the fact that $(G_{th} - R)_e$ is the number of electrons transferred from all the traps in the bandgap to the conduction band per unit time per unit volume. Obviously, we can express $(G_{th} - R)_e$ as

$$(G_{th} - R)_e = R_{ee} - R_{ec} = \int_{-E_g/2}^{E_g/2} (r_{ee} - r_{ec})\, d(E - E_{Fi}) \tag{1.71}$$

Similarly, $(G_{th} - R)_h$ can be expressed as

$$(G_{th} - R)_h = R_{he} - R_{hc} = \int_{-E_g/2}^{E_g/2} (r_{he} - r_{hc})\, d(E - E_{Fi}) \tag{1.72}$$

In a dc steady state, however, these two rates are equal, as discussed in Section 1.1.4. According to (1.71) and (1.72), this equality implies $r_{ee} - r_{ec} = r_{he} - r_{hc}$. Substituting (1.67) through (1.70) into the latter equation, solving for n_t, substituting the solution back into (1.67) and (1.68), and finally, using the latter two in (1.71), we obtain the net generation rate in steady state as

$$G_{th} - R = (n_i^2 - np) \int_{-E_g/2}^{E_g/2} \frac{v_{th} D_t d(E - E_{Fi})}{\dfrac{n}{\sigma_p} + \dfrac{p}{\sigma_n} + \dfrac{n_i}{\sigma_p} \exp\left(\dfrac{E - E_{Fi}}{kT}\right) + \dfrac{n_i}{\sigma_n} \exp\left(-\dfrac{E - E_{Fi}}{kT}\right)} \tag{1.73}$$

Note that the same result would be obtained by substituting the solution for n_t back into (1.69) and (1.70) and then using the latter two in (1.72).

Since the integral in (1.73) is always non-negative, the polarity of the net generation rate is determined by the polarity of the prefactor $(n_i^2 - np)$. In thermal equilibrium, the law of mass action implies $np = n_i^2$, hence, $G_{th} - R = 0$, as expected. In a nonequilibrium state, however, the law of mass action does not hold, that is, $np \neq n_i^2$. If $np < n_i^2$, $G_{th} - R$ becomes positive, which indicates a dominance of thermal generation over recombination. In a way, the semiconductor tries to raise the carrier concentrations back to their equilibrium levels by suppressing recombination. In the opposite case of $np > n_i^2$, $G_{th} - R$ becomes negative, which indicates an enhanced recombination. Obviously, the semiconductor again reacts in the direction to restore equilibrium.

Finally, note that the conclusions reached in this subsection, including (1.73), are valid everywhere in a silicon device operating in a dc steady state. They can be applied to the nonequilibrium analysis of space-charged regions and bulk regions equally well.

1.3.3 Analysis of a Bulk Region in Low-Level Injection

Continuity Equations

The continuity equations for a dc steady state contain $G_{th} - R$, and the latter is given by the rather awesome formula of (1.73). Fortunately, under low-level injection conditions, this equation of net generation rate—hence, the continuity equations—can be simplified to such an extent that a simple, yet accurate, analytical treatment of bulk regions becomes possible. We will now derive this simple form of the continuity equations.

Consider an n-type bulk ($N > 0$) for which the conditions of low-level injection imply $n \cong n_o$ for the majority-carrier electrons [see (1.65)]. For the minority-carrier holes, we can write $p = p_o + p'$ from (1.62). Using these two equations, the prefactor $n_i^2 - np$ of the integral in (1.73) can be written as

$$n_i^2 - np = n_i^2 - n_o(p_o + p') = n_i^2 - n_op_o - p'n_o$$

Considering $n_op_o = n_i^2$, we can turn this equation into $n_i^2 - np = n_op'$. Using the latter in (1.73), assuming σ_n and σ_p to be equal and energy independent, neglecting p in comparison with n in the denominator of the integrand as a consequence of low-level injection, and finally, substituting (1.41) for n because $n \cong n_o$, we obtain the following equation for the net generation rate

$$G_{th} - R = -v_{th}\sigma p' \int_{-E_g/2}^{E_g/2} W(E)D_t \, d(E - E_{Fi}) \tag{1.74}$$

where

$$\sigma \equiv \sigma_n = \sigma_p,$$

$$W(E) \equiv \frac{1}{1 + \exp\left[\dfrac{E - E_{Fi} - (E_F - E_{Fi})}{kT}\right] + \exp\left[\dfrac{-(E - E_{Fi}) - (E_F - E_{Fi})}{kT}\right]}$$

As understood from the plot given in Figure 1.17, $W(E)$ can be closely approximated by

$$W(E) = 0 \qquad \text{for } E - E_{Fi} < -(E_F - E_{Fi}) \text{ and } E - E_{Fi} > E_F - E_{Fi} \tag{1.75a}$$

$$W(E) = 1 \qquad \text{for } -(E_F - E_{Fi}) < E - E_{Fi} < E_F - E_{Fi} \tag{1.75b}$$

which enables us to simplify (1.74) into

$$G_{th} - R = -v_{th}\sigma N_t p' \tag{1.76}$$

where

$$N_t \equiv \int_{E_{Fi} - E_F}^{E_F - E_{Fi}} D_t d(E - E_{Fi}) \tag{1.77}$$

is the effective trap density per unit volume. It is obvious from the limits of the integral in (1.77) and from Figure 1.17 that a trap can contribute to the generation and recombination process inside a bulk region of low-level injection only if its energy level lies within an energy range of width $2(E_F - E_{Fi})$ centered around midband. According to (1.41) and (1.52), this range of effective traps is determined solely by the doping concentration N. Knowing the latter and the trap density D_t, theoretically, one can calculate N_t from (1.77). In practice, however, neither the D_t of (1.77) nor the σ of (1.76) is easily extractable. For this reason, (1.76) is usually expressed as

$$G_{th} - R = -\frac{p'}{\tau_p} \tag{1.78}$$

where the so-called *minority-carrier* (hole) *lifetime*

$$\tau_p \equiv \frac{1}{v_{th}\sigma N_t} \tag{1.79}$$

solely represents the trapping-related structural parameters. Several methods are available for determining the lifetime experimentally [10–13].

Repeating the preceding analysis for a p-type bulk operating under low-level injection conditions in a dc steady state, we obtain similar results. In this type of bulk, E_F lies in the lower half bandgap as discussed in Section 1.2.1. For this reason the limits of the integral in (1.77) must be transposed. As to the counterparts of (1.78) and (1.79), we obtain

$$G_{th} - R = -\frac{n'}{\tau_n} \tag{1.80}$$

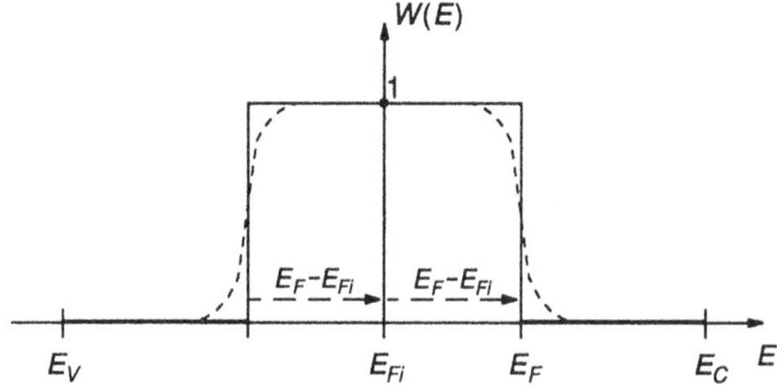

Figure 1.17 The energy window defined by the function $W(E)$ for the traps that can contribute to the net generation rate in an n-type bulk region operating under low-level injection conditions.

and

$$\tau_n \equiv \frac{1}{v_{th}\sigma N_t}$$

Now we can construct the low-level injection form of the two continuity equations simply by substituting (1.78) or (1.80) into (1.25) and (1.26), and ignoring g. The results are as follows:

n-*type bulk:*

$$\frac{dJ_p}{dx} = -\frac{q}{\tau_p}p' \tag{1.81}$$

$$\frac{dJ_n}{dx} = \frac{q}{\tau_p}p' \tag{1.82}$$

p-*type bulk:*

$$\frac{dJ_p}{dx} = -\frac{q}{\tau_n}n' \tag{1.83}$$

$$\frac{dJ_n}{dx} = \frac{q}{\tau_n}n' \tag{1.84}$$

Minority Carriers

Having simplified the continuity equations for low-level injection conditions, we are now ready to analyze a bulk region operating under these conditions. We derive the minority-carrier concentration profile first.

The total current density in a semiconductor can be expressed as

$$J = J_n + J_{p(dr)} + J_{p(dif)} \tag{1.85}$$

where

$$J_n = q\mu_n n\mathscr{E} + q\frac{kT}{q}\mu_n\frac{dn}{dx}$$

is the electron current density, and

$$J_{p(dr)} \equiv q\mu_p p\mathscr{E}$$

and

$$J_{p(dif)} \equiv -q\frac{kT}{q}\mu_p\frac{dp}{dx}$$

denote the hole drift current density and hole diffusion current density, respectively. Now suppose that an n-type bulk region is perturbed with low-level injection. Although the current density in the region is generally described by (1.85), the low-level injection condition enables us to simplify this equation considerably. First, note that the region is still n-type, and $n \gg p$ as a result of low-level injection. Therefore, the hole (minority-carrier) drift current is negligible in comparison with the electron (majority-carrier) drift current because μ_n and μ_p are of the same order of magnitude. Therefore,

$$J \cong J_n + J_{p(dif)} \tag{1.86}$$

Now, assume that the excitation causing the nonequilibrium state is time invariant, and that a steady state has been reached, which implies from (1.27) a position-independent J. Substituting (1.86) into (1.27), we obtain

$$\frac{dJ_n}{dx} = -\frac{dJ_{p(dif)}}{dx}$$

Using this in the continuity equation (1.82), replacing $J_{p(dif)}$ with $-qD_p\, dp/dx$, and taking into account that p equals $p_o + p'$ and p_o is practically position independent because a bulk region is uniformly or near-uniformly doped, we arrive at the equation

$$\frac{d^2p'}{dx^2} = \frac{p'}{L_p^2} \tag{1.87}$$

where

$$L_p \equiv \sqrt{D_p\tau_p} \tag{1.88}$$

is called the *minority-carrier* (hole) *diffusion length*. We now have a differential equation describing the position dependence of the minority-carrier excess concentration in an n-type bulk operating under low-level injection conditions in steady state.

Let us consider the solutions of (1.87) for the two extreme cases of very long and very short bulk regions. In the former case, in which, the bulk length L is much larger than L_p, the solution is given by

$$p'(x) = p'(0)\exp(-x/L_p), \qquad \text{for } L \gg L_p \tag{1.89}$$

where $p'(0)$ is the boundary value of the hole excess concentration at $x = 0$. In the other extreme case, that of a very short bulk region, the solution is given by

$$p'(x) = p'(0) + \frac{p'(L) - p'(0)}{L}x, \qquad \text{for } L \ll L_p \tag{1.90}$$

where $p'(L)$ is the boundary value of the hole excess concentration at $x = L$.

According to (1.89), if minority carriers have an excess concentration at the boundary of a "long" bulk region of length $L \gg L_p$, this excess concentration varies exponentially with position in the bulk, and vanishes in a distance of a few L_p away from the boundary. Deep in the bulk, therefore, the minority-carrier concentration reaches its equilibrium level n_i^2/N as shown in Figure 1.18(a). The physical mechanism that reduces the excess concentration is recombination for $p'(0) > 0$ and generation for $p'(0) < 0$. In the former case, where minority carriers are continuously injected to the bulk at $x = 0$ to maintain a positive $p'(0)$, these

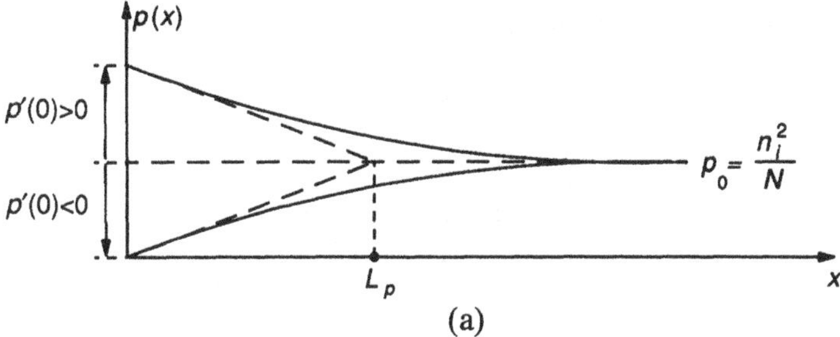

(a)

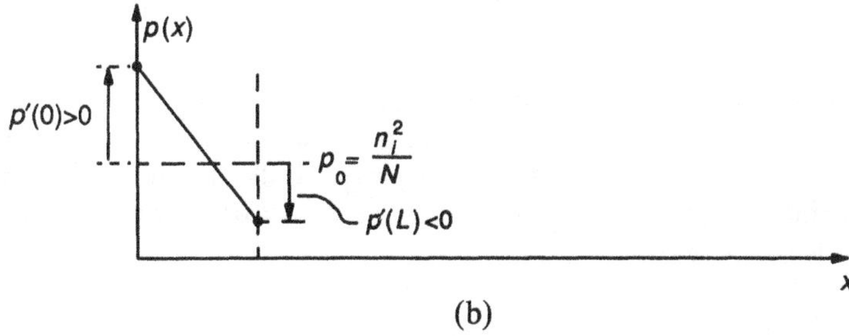

(b)

Figure 1.18 The spatial variation of the minority-carrier concentration in two extreme cases of bulk geometry: (a) The bulk length is much greater than the diffusion length L_p. Notice that the carrier concentration varies exponentially to approach the equilibrium value n_i^2/N_A. (b) A very short bulk of length $L \ll L_p$. Notice the linear variation between the two boundary values.

carriers diffuse into the bulk, and simultaneously recombine with majority carriers. Deep in the bulk, where the excess concentration vanishes, recombination is balanced by generation, and thus equilibrium conditions are restored.

The case of $p'(0) < 0$ corresponds to a continuous extraction of minority carriers at $x = 0$. This results in a deficiency of minority carriers, thus decreasing the rate of recombination below the equilibrium level at and near the surface. The minority carriers thermally generated in this region diffuse toward the boundary to act as the supply for the extraction process.

According to (1.90), the minority-carrier excess concentration in a "short" bulk of length $L \ll L_p$, has a linear profile varying between the two boundary values. An example is given in Figure 1.18(b). Because minority carriers flow mainly by diffusion, a linear concentration profile implies a position-independent minority-carrier current, which, in turn, implies a negligible generation and recombination process as understood from continuity equation (1.25). In fact, generation and recombination processes do exist, but the region is so short that their effect on the minority-carrier concentration is negligibly small.

Repeating the foregoing analysis for a p-type ($N < 0$) bulk region operating under low-level injection conditions in a dc steady state, we reach similar conclusions. But, of course, electrons are the minority carriers in a p-type bulk. The main equations, as expressed in appropriate notation are as follows:

$$\frac{d^2n'}{dx^2} = \frac{n'}{L_n^2} \tag{1.91}$$

$$L_n \equiv \sqrt{D_n \tau_n} \tag{1.92}$$

$$n'(x) = n'(0)\exp(-x/L_n), \qquad \text{for } L \gg L_n \tag{1.93}$$

$$n'(x) = n'(0) + \frac{n'(L) - n'(0)}{L}x, \qquad \text{for } L \ll L_n \tag{1.94}$$

Once the minority-carrier concentration profile is in hand, one can easily obtain the minority-carrier current density profile from $J_p = -qD_p(dp'/dx)$ for an n-type bulk, and from $J_n = qD_n(dn'/dx)$ for a p-type bulk. For the electric field and majority-carrier concentration profiles, we need to further expand the foregoing analysis, as is done next.

Quasineutrality and Majority Carriers

Let us return to (1.85) and the subsequent equations, and take the following analytical steps: For dn/dx, substitute

$$\frac{dn}{dx} = \frac{dp}{dx} - \frac{\varepsilon}{q}\frac{d^2\mathscr{E}}{dx^2}$$

which is derived by differentiating Poisson's equation (1.5) assuming $dN/dx \cong 0$ as required by the neutrality of the bulk in equilibrium. Next, ignore $J_{p(dr)}$, and replace the majority-carrier concentration n with N in J_n, as implied by the low-level injection condition. Then make use of definition (1.58) of the Debye length, and replace dp/dx with dp'/dx because $p = p_o + p'$, and $p_o = n_i^2/N$ can be assumed to be position independent in a bulk. With these steps, we can turn (1.85) into

$$L_D^2\frac{d^2\mathscr{E}}{dx^2} - \mathscr{E} + \frac{J}{q\mu_n N} + \frac{kT}{q}\frac{1}{N}\left(\frac{\mu_p}{\mu_n} - 1\right)\frac{dp'}{dx} = 0$$

According to (1.89) and (1.90), the last term on the left-hand side of this equation equals

$$-\frac{kT}{q}\frac{1}{L_p}\left(\frac{\mu_p}{\mu_n} - 1\right)\frac{p'(0)}{N}\exp(-x/L_p), \qquad \text{for } L \gg L_p$$

or

$$-\frac{kT}{q}\frac{1}{L}\left(\frac{\mu_p}{\mu_n} - 1\right)\frac{p'(0) - p'(L)}{N}, \qquad \text{for } L \ll L_p$$

This term, however, is negligibly small in both cases because the excess concentrations are much smaller than the doping concentration due to the low-level injection condition. This leaves us with

$$L_D^2\frac{d^2\mathscr{E}}{dx^2} - \mathscr{E} + \frac{J}{q\mu_n N} = 0$$

Considering the fact that L_D is much smaller than the length of any practical bulk region, the solution of this equation is obtained as

$$\mathscr{E}(x) - \frac{J}{q\mu_n N} = \left[\mathscr{E}(0) - \frac{J}{q\mu_n N}\right]\exp(-x/L_D) \qquad (1.95)$$

Since N and J are position independent, the position dependence of $\mathscr{E}$ stems only from the exponential factor on the right-hand side of this equation. However, L_D is much smaller than the length of a typical bulk. Therefore, even if the boundary value $\mathscr{E}(0)$ of the field is different from $J/q\mu_n N$, this difference is bound to vanish in a very short distance away from the boundary, leaving the rest of the bulk with a position-independent field:

$$\mathscr{E} \cong \frac{J}{q\mu_n N} \tag{1.96}$$

According to Poisson's equation, a position-independent field implies $\rho = 0$. We, therefore, conclude that a bulk region operating under low-level injection conditions in a dc steady state retains its neutrality. This is what we call *quasineutrality.* Obviously, $\rho = 0$ generally implies

$$p - n + N = p_o + p' - (n_o + n') + N = 0$$

Since $p_o - n_o + N = 0$ in a bulk region, this equation can be reduced to

$$n' = p' \tag{1.97}$$

and accordingly, the majority-carrier concentration can be expressed as

$$n = N + p' \tag{1.98}$$

In a p-type ($N < 0$) bulk, (1.97) remains valid but (1.96) and (1.98) are replaced by

$$\mathscr{E} \cong \frac{J}{q\mu_p |N|} \tag{1.99}$$

and

$$p = |N| + n' \tag{1.100}$$

respectively.

Boundary Conditions at an Ohmic Contact

Most of the analytical results of the foregoing analysis of bulk regions are expressed in terms of the boundary value of the excess carrier concentration. Generally, this value is determined collectively by the bulk under analysis and the neighboring region extending beyond the boundary. The only exception of practical significance is the case of a boundary formed by an *ohmic contact,* where the excess concentration vanishes regardless of the properties of the neighboring region. Such a boundary is created wherever a bulk is contacted with a conductor for the purpose of establishing a terminal of very low resistance. The microstructure of this contact gives rise to an infinitely large effective trap density N_t in an infinitesimally narrow bulk region adjacent to the boundary. As understood from (1.79), (1.81), and (1.82) for an n-type bulk, $N_t \to \infty$ implies $p' \to 0$ and therefore $n' \to 0$ in this boundary layer because, otherwise, the boundary values of J_p and J_n would be infinitely large. A similar conclusion can be easily reached for the boundary of a p-type bulk terminated by an ohmic contact.

1.3.4 Extending the Fermi Formalism to Nonequilibrium

Since the law of mass action (1.39) is invalid in nonequilibrium, the nonequilibrium carrier concentrations n and p cannot be described by a single Fermi level as is done with (1.41) and (1.42) for equilibrium. Yet, the concept of the Fermi level is so useful that we tend to apply it also to nonequilibrium. To do this, we need to define two separate *quasi-Fermi levels* for electrons and holes. Using the notation E_{Fn} and E_{Fp} for these, we can express n and p as follows:

$$n = n_i \exp\left(\frac{E_{Fn} - E_{Fi}}{kT}\right) \tag{1.101}$$

$$p = n_i \exp\left(-\frac{E_{Fp} - E_{Fi}}{kT}\right) \tag{1.102}$$

Note the formal similarity between these two equations and (1.41) and (1.42).

Naturally, we also can define two *quasi-Fermi potentials* by extending definition (1.43) of the equilibrium Fermi potential; that is,

$$\phi_{Fn} \equiv -\frac{E_{Fn} - E_{Fi}}{q} \tag{1.103}$$

$$\phi_{Fp} \equiv -\frac{E_{Fp} - E_{Fi}}{q} \tag{1.104}$$

Combining (1.103) with (1.101), and (1.104) with (1.102), we finally obtain

$$n = n_i \exp\left(-\frac{q}{kT}\phi_{Fn}\right) \tag{1.105}$$

$$p = n_i \exp\left(\frac{q}{kT}\phi_{Fp}\right) \tag{1.106}$$

as the nonequilibrium counterparts of (1.44) and (1.45), respectively.

An alternative form of the current density equations, expressed in terms of the two quasi-Fermi levels, is frequently used in device analysis. To derive this form, substitute into (1.13) and (1.18) the spatial derivatives of p and n from (1.102) and (1.101) together with dE_{Fi}/qdx for $\mathscr{E}$ [from (1.31)], and use the Einstein relationships (1.16) and (1.21) for D_p and D_n. The result will be

$$J_p = \mu_p p \frac{dE_{Fp}}{dx} \tag{1.107}$$

$$J_n = \mu_n n \frac{dE_{Fn}}{dx} \tag{1.108}$$

PROBLEMS

1.1 We know from Section 1.2.2 that a uniform doping concentration perfectly satisfies the necessary condition (1.59) of neutrality in equilibrium but, in order for a region of uniform doping concentration to be neutral in its entirety, the boundary field $\mathscr{E}(0)$ must also be zero [see (1.61) and (1.60).] If $\mathscr{E}(0) \neq 0$, a space-charge region develops next to the boundary. Now, consider a p-type silicon bar doped to a net concentration $N = -N_A$ and subjected to a *positive* boundary field $\mathscr{E}(0)$ in a thermal equilibrium state at $T = 300$K, as shown in Figure P1.1(a). Since $\mathscr{E}(0) > 0$, the positively charged holes are repelled from the boundary, whereas the negatively charged electrons are attracted to it.

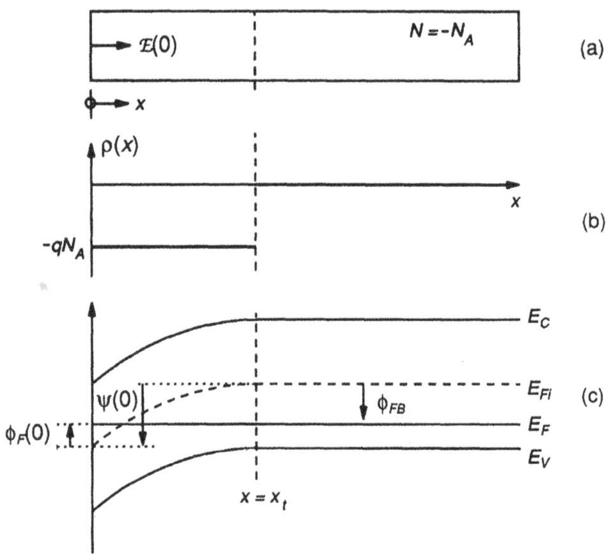

Figure P1.1

This results in $p_o < N_A$ and $n_o > n_i^2/N_A$ in the surface space-charge region. If $\mathscr{E}(0)$ is not too strong to make n_o comparable to N_A, both p_o and n_o will be much smaller than N_A in most of this region. This condition, called *depletion*, enables us to approximate the charge-density profile with the *box function* shown in Figure P1.1(b), where x_t denotes the width of the depletion region.

(a) Using Poisson's equation, determine the electric field profile $\mathscr{E}(x)$ inside this surface depletion region in terms of q, N_A, ϵ, and $\mathscr{E}(0)$. Considering the fact that $\mathscr{E} = 0$ for $x \geq x_t$, sketch the profile.

(b) Find an expression for x_t in terms of q, N_A, ϵ, and $\mathscr{E}(0)$. Does the depletion region expand or contract with an increasing $\mathscr{E}(0)$? Calculate x_t for $N_A = 10^{15}$ cm^{-3} and $\mathscr{E}(0) = 10^4$ V/cm. (*Answer:* $x_t = 6.5 \times 10^{-5}$ cm $= 0.65$ μm.)

(c) Assuming $\psi = 0$ in the bulk region extending beyond $x = x_t$, determine from (1.6) the electrostatic potential profile $\psi(x)$ in the depletion region in terms of q, N_A, ϵ, and $\mathscr{E}(0)$, and sketch the profile. Give an expression for the so-called "surface potential" $\psi(0)$ (ψ at $x = 0$) in terms of the same parameters. Calculate it for the numerical values given in part (b). (*Answer:* $\psi(0) = 0.325$ V.)

(d) The energy-band profile of the silicon bar is given in Figure P1.1(c). Calculate n_o, p_o, and the Fermi potential ϕ_{FB} in the bulk region. Why are the bands straight in the bulk? (*Answer:* $n_o = 1.21 \times 10^5$ cm^{-3}, $p_o = 10^{15}$ cm^{-3}, $\phi_{FB} = 0.295$ V.)

(e) Calculate the carrier concentrations $n_o(0)$ and $p_o(0)$ at $x = 0$, and compare them with N_A to see whether the boundary is indeed depleted. (*Answer:* $n_o = 3.52 \times 10^{10}$ cm^{-3}, $p_o = 3.44 \times 10^9$ cm^{-3}.)

(f) For which particular value of $\mathscr{E}(0)$ does $n(0)$ equal $p(0)$? (*Answer:* $\mathscr{E}(0) = 9.53 \times 10^3$ V/cm.)

(g) If you increase $\mathscr{E}(0)$ further than what you calculated in part (f), $n(0)$ exceeds $p(0)$. For a particular value of $\mathscr{E}(0)$, $n(0)$ equals N_A. This condition, beyond which the assumption of a depleted surface space-charge region is no more justifiable, is referred to as the *threshold* condition in MOS terminology. Calculate the boundary field that causes threshold in the present example. (*Answer:* $\mathscr{E}(0) = 1.35 \times 10^4$ V/cm.)

(h) If a boundary located at, say, $x = 0$, in a dielectric medium contains a two-dimensional (sheet) charge of density Q [C/cm^2], then, the displacement vector must have a discontinuity of $D(0^+) - D(0^-) = Q$ at that boundary, as can be shown from (1.1). We will invoke this rule in Chapter 3 in analyzing the MOSFET. As a primer, suppose that the region lying to the left of $x = 0$ in Figure P1.1(a) is an MOS gate insulator made of SiO_2 whose dielectric constant is $\epsilon_{ox} = 3.45 \times 10^{-13}$ F/cm, and that a sheet charge of density $Q_{ss} = 10^{-8}$ C/cm^2 (called *oxide fixed charge* in MOS terminology) resides at $x = 0$. Calculate the field $\mathscr{E}(0^-)$ on the oxide side of the boundary for the threshold condition. (*Answer:* $\mathscr{E}(0^-) = 1.17 \times 10^4$ V/cm.)

1.2 Shown in Figure P1.2(a) is the doping concentration profile around a *pn* junction. Traditionally, the space-charge region of a *pn* junction is assumed to be depleted of carriers between a point x_1 on the *n* side and a point x_2 on the *p* side. This leads to the charge density and electric field profiles shown in Figure P1.2(b,c), respectively.

(a) Show that the slope of the field profile is given by qN_D/ϵ on the *n* side and by $-qN_A/\epsilon$ on the *p* side, and that the triangular area bounded by the field profile equals the electrostatic potential difference $\Delta\psi \equiv \psi(x_1) - \psi(x_2)$.

(b) Relying on the geometric features of the field profile, you can easily write the following three equations:

$$\frac{\mathscr{E}(0)}{-x_1} = \frac{q}{\epsilon}N_D$$

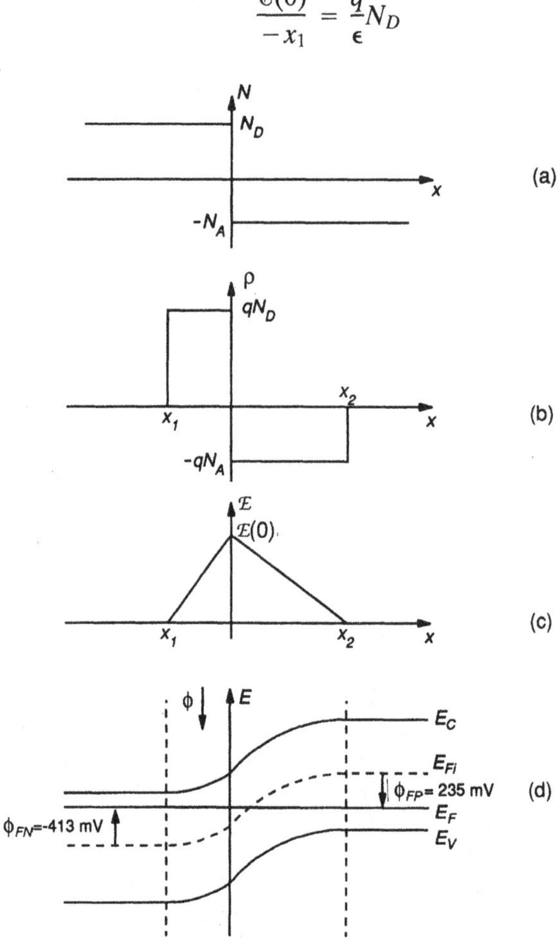

Figure P1.2

$$\frac{\mathscr{E}(0)}{x_2} = \frac{q}{\epsilon} N_A$$

$$\Delta\psi = \frac{x_2 - x_1}{2} \mathscr{E}(0)$$

Verify these equations and then use them to construct expressions for x_1 and x_2 in terms of q, ϵ, N_A, N_D, and $\Delta\psi$. Also find an expression for $W_d \equiv x_2 - x_1$, which is the total width of the depletion region.

(c) Figure P1.2(d) gives the energy-band diagram of the junction in thermal equilibrium. Determine N_A, N_D, and $\Delta\psi$ from this diagram. Using them in the expressions found in part (b), calculate x_1, x_2, and W_t. Notice that if $N_A \neq N_D$, a larger part of the depletion region resides on the more lightly doped side of the junction. (*Answer:* $x_1 = -2.90 \times 10^{-7}$ cm, $x_2 = 2.90 \times 10^{-4}$ cm $= 2.90$ μm, $W_d \cong x_2 = 2.90$ μm.)

(d) The potential difference defined in part (a) as $\Delta\psi$ is called the *built-in voltage* or *barrier potential* of the *pn* junction, and is denoted by V_b in thermal equilibrium. When the *p* side of the junction is biased with an external voltage V with respect to the *n* side, this potential difference becomes $\Delta\psi = V_b - V$. Review the results of part (b) using this equation for $\Delta\psi$. Examine the effect of V on x_1, x_2, W_d, and $\mathscr{E}(0)$ for $V > 0$ (forward bias) and for $V < 0$ (reverse bias).

1.3 The derivation of the current density equations, (1.13) and (1.18), was implicitly based on an assumption of a homogeneous semiconductor material. In this problem, you are expected to modify these equations for the more general case of a heterogenous semiconductor, in which both E_g and κ are arbitrary functions of position. Fortunately, you don't need to start from scratch. For the holes, start with (1.9), whose right-hand side contains $q\mathscr{E}$ as the force acting on a hole in a homogeneous semiconductor. More generally, this force can be expressed as

$$F_p = \frac{dE_V}{dx}$$

where E_V, being the uppermost valence band level, represents hole energy.

(a) Use this more general equation of F_p in (1.9) for $q\mathscr{E}$, and then proceed with the aid of Figure 1.11(a) to the generalized hole current density equation

$$J_p = q\mu_p p\left[\mathscr{E} - \frac{d(E_g + \kappa)}{q\,dx}\right] - qD_p\frac{dp}{dx}$$

Elaborate on the effect of the heterogeneity on hole current density.

(b) Starting with

$$F_n = -\frac{dE_C}{dx}$$

repeat part (a) for the electron current density equation.

(c) Having obtained the generalized forms of the two current density equations, you now have an opportunity to verify the E_g-dependence of n_i as defined by (1.40). To this end, rewrite the generalized forms of J_p and J_n for thermal equilibrium [like (1.37) and (1.38)], and then manipulate them to show

$$n_i = M \exp\left(-\frac{E_g}{2kT}\right)$$

where M is a constant.

1.4 For a uniformly doped region, the general equation (1.49) of equilibrium can be reduced to a first-order differential equation of tremendous analytical utility. The reduction is done by multiplying both sides of (1.49) by $2d\phi_F/dx$ and then rewriting the left-hand side as $d(d\phi_F/dx)^2/dx$.

(a) Reduce the equation, and show that the resulting first-order equation yields

$$\mathscr{E}^2(x) = 2\int_{\phi_F(0)}^{\phi_F(x)} F(\xi)d\xi + \mathscr{E}^2(0)$$

where

$$F(\xi) \equiv \frac{q}{\epsilon}\left[n_i \exp\left(\frac{q}{kT}\xi\right) - n_i \exp\left(-\frac{q}{kT}\xi\right) + N\right]$$

is the right-hand side of (1.49) as written in terms of the integration variable ξ.

(b) Now, suppose that a uniformly doped region is subject to an electric field $\mathscr{E}(0)$ at $x = 0$. Unless $\mathscr{E}(0) = 0$, a space-charge region will develop at the surface. In Problem 1.1 you analyzed this region assuming depletion. But you also know that the condition of depletion prevails only for a specific range of $\mathscr{E}(0)$ values. Without imposing such a restriction on $\mathscr{E}(0)$, you can quite generally relate $\phi_F(0)$ to $\mathscr{E}(0)$ using the equation given in part (a). The relationship for a p-type semiconductor ($N = -N_A$) is as follows:

$$\mathscr{E}(0) = \pm\left[2\frac{q}{\epsilon}\left(N_A[\phi_{FB} - \phi_F(0)] + \frac{kT}{q}n_i\left\{\exp\left[\frac{q}{kT}\phi_F(0)\right] + \exp\left[-\frac{q}{kT}\phi_F(0)\right]\right\}\right.\right.$$
$$\left.\left. - \frac{kT}{q}n_i\left[\exp\left(\frac{q}{kT}\phi_{FB}\right) + \exp\left(-\frac{q}{kT}\phi_{FB}\right)\right]\right)\right]^{1/2}$$

where ϕ_{FB} is the Fermi potential in the neutral bulk. Derive this equation.

(c) Assuming a silicon bar of $N_A = 1.23 \times 10^{15}\ \text{cm}^{-3}$ at $T = 300\text{K}$, calculate $\mathscr{E}(0)$ for the following values of $\phi_F(0)$: -0.50V, -0.45V, -0.40V, -0.35V, -0.30V, -0.20V, -0.10V, 0V, 0.10V, 0.20V, 0.30V, 0.35V, 0.40V, 0.45V, and 0.50V. The value of $\mathscr{E}(0)$ must be positive for $\phi_F(0) < \phi_{FB}$ and negative for $\phi_F(0) > \phi_{FB}$; why? Now, plot $\phi_F(0)$ as a function of $\mathscr{E}(0)$.

(d) A surface space-charge region can be in one of five possible conditions. For a region of $N = -N_A$ in equilibrium, these conditions are defined as: (1) Accumulation: $p(0) > N_A$. (2) Flatband: $p(0) = N_A$. (3) Depletion: $p(0) < N_A$ and $n(0) < N_A$. (4) Threshold: $n(0) = N_A$. (5) Inversion: $n(0) > N_A$. You already know the third and fourth of these conditions. Now you have a chance to learn the rest. First persuade yourself that these definitions are tantamount to (1) $\phi_F(0) > \phi_{FB}$ for accumulation, (2) $\phi_F(0) = \phi_{FB}$ for flatband, (3) $-\phi_{FB} < \phi_F(0) < \phi_{FB}$ for depletion, (4) $\phi_F(0) = -\phi_{FB}$ for threshold, and (5) $\phi_F(0) < -\phi_{FB}$ for inversion. Now return to the plot of part (c) and mark the flatband and threshold conditions so that the remaining three conditions are properly demarcated on the plot. How does the sensitivity of $\phi_F(0)$ to $\mathscr{E}(0)$ change as you scan the semiconductor from accumulation to inversion? Show that the relative sensitivity $|d\phi_F(0)/[d\mathscr{E}(0)/\mathscr{E}(0)]|$ approaches $2kT/q$ in the limit of very strong inversion or accumulation. Relying on this conclusion of weak sensitivity, we will assume a field-independent $\phi_F(0)$ in strong inversion when we model the MOSFET operation in Chapter 3.

(e) Sketch the energy-band diagram separately for the five possible conditions of the surface space-charge region.

1.5 A very long silicon bulk region is doped uniformly to $N = N_D = 10^{16}\ \text{cm}^{-3}$. The trap density D_t equals $5 \times 10^{15}\ \text{cm}^{-2}\cdot\text{eV}^{-1}$ for $0.1\ \text{eV} \le E - E_{Fi} \le 0.2\ \text{eV}$ and $E - E_{Fi} \ge 0.45$ eV, but is negligibly small in the rest of the bandgap. The capture cross section of these traps is $10^{-15}\ \text{cm}^{-3}$, and the thermal velocity of carriers is 10^7 cm/s. Assume $T = 300\text{K}$ for the entire problem.

(a) Calculate the minority-carrier lifetime. (*Answer:* 0.2 μs.)

(b) Calculate the minority-carrier diffusion length. (*Answer:* 15 μm.)

(c) The bulk is perturbed with $p'(0) = 10^9\ \text{cm}^{-3}$. Calculate and plot on a semilog coordinate system the profiles $p(x)$ and $n(x)$ for the range $0 \le x \le 225$ μm. These plots will give you a fair idea as to how the minority-

and majority-carrier concentrations are affected by a typical low-level injection condition.

(d) Calculate and plot J_p and the diffusion component of J_n as functions of position.

(e) Assuming $J = 1.3\ \mu\text{A/cm}^2$, calculate and plot the drift component of J_n as a function of x.

(f) Now, relying on the plots drawn in parts (d) and (e), discuss in physical terms the roles assumed by holes and electrons in current transport along the region.

(g) Since the region remains in a quasineutral state, the E_C, E_V, and E_{Fi} levels of the energy-band diagram are still flat. But now, the holes and electrons must generally have two separate quasi-Fermi levels, E_{Fn} and E_{Fp}. How do these levels vary with position?

1.6 Shown in Figure P1.6 are the geometric features of a semiconductor resistor, which comprises a p-type or n-type semiconductor bar and two terminals connected to ohmic-contact surfaces at $x = 0$ and $x = L$. Application of a voltage V between these terminals results in a one-dimensional current that flows along x across a cross-sectional area of width W and height H. In the most general case, the doping concentration N varies with all three position variables x, y, and z.

(a) Beginning with the fact that the semiconductor is neutral in equilibrium, and assuming that $N = -N_A(x,y,z)$, show that the potential difference $V_b \equiv \psi(0) - \psi(L)$ in equilibrium is given by

$$V_b = \frac{kT}{q} \ln \frac{N_A(L)}{N_A(0)}$$

(b) Due to the presence of ohmic contacts at the two terminal surfaces, no excess concentration can develop in this structure in nonequilibrium. Taking this fact into account, write a general expression for the total current density using (1.13) and (1.18). Then show from this expression that the

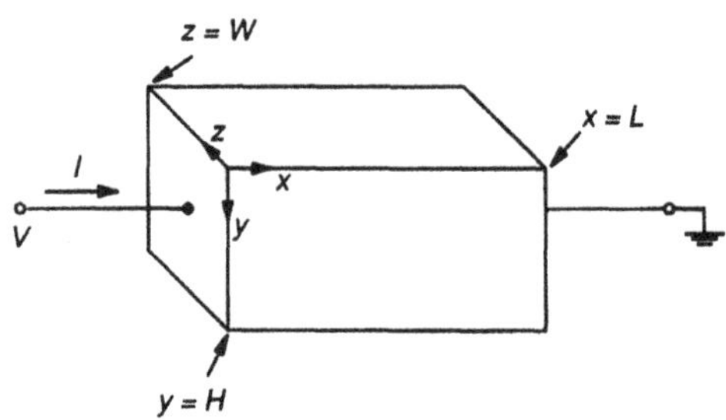

Figure P1.6

minority-carrier current density components are negligible in comparison with their majority-carrier counterparts. Ignoring these negligible components, show that the boundary-to-boundary potential difference in nonequilibrium can be described as

$$\psi(0) - \psi(L) = J\int_0^L \frac{dx}{q\mu_p N_A(x, y, z)} + V_b$$

(c) The potential difference $\psi(0) - \psi(L)$ in nonequilibrium equals $V_b + V$, where V is the external bias related to the current I by $V = RI$. Show that, if N_A varies only with x, the resistance R can be expressed as

$$R = \frac{1}{HW}\int_0^L \frac{dx}{q\mu_p N_A(x)}$$

(d) Find an expression for R in the case of N_A being a function of z only.

(e) In integrated *diffusion* resistors, the doping concentration is a function of y only. The structural features along this direction are determined by the technologist, and are presented to the circuit designer as a single parameter called *sheet resistance* (R_S). A circuit designer, who has full freedom in setting device geometries in the x and z directions but not in the y direction relies on R_S to determine the length L and width W of the resistor so that the desired resistance can be obtained. In other words, R is determined by W, L, and R_S only. Find an expression for R_S in terms of y-dependent structural features.

1.7 The last topic we discussed in Section 1.3.3 was the boundary conditions at an ohmic contact. We based the discussion on the assumption that such a contact involved an infinitely large effective trap density in an infinitesimally narrow boundary layer, which led us to the conclusion that excess concentrations should vanish at the boundary. We now relax this rather harsh assumption and consider the case in which N_t in the narrow boundary layer is *not* infinitely large. Shown in Figure P1.7 is the boundary layer under consideration. It belongs to an n-type bulk operating under low-level injection conditions, and it contains N_t effective traps per unit volume in between $x = 0$ and $x = 0^+$. This layer is so narrow that excess concentrations inside the layer are virtually independent of the position variable, i.e., $p' = p'(0)$ and $n' = n'(0)$ for $0 \leq x \leq 0^+$. Also note that the metallic region lying to the left of the layer cannot support a hole current, hence $J_p(0) = 0$.

(a) Using (1.81) and (1.82), derive the relationships

$$J_p(0^+) = -qsp'(0)$$

$$J_n(0^+) = J + qsp'(0)$$

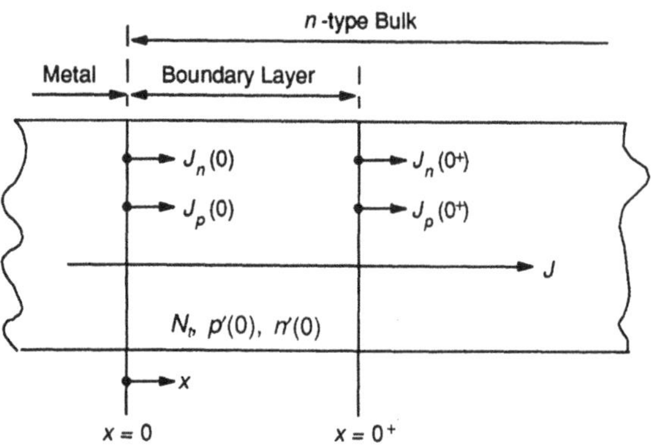

Figure P1.7

and write an equation of definition for the parameter s, which represents the *surface recombination velocity*. Observe that $p'(0)$ and $n'(0)$ can vanish only if s is infinite.

(b) Repeat part (a) for a p-type bulk.

(c) Now consider an n-type bulk of length $L \ll L_p$, which has an ohmic contact of surface recombination velocity s at $x = L$, and operates under low-level injection conditions with $p'(0)$ at $x = 0$. Show that

$$J_p = \frac{qp'(0)}{\dfrac{L}{D_p} + \dfrac{1}{s}}$$

1.8 Shown in Figure P1.8(a) is the schematic of a biased pn junction structure. You are expected to model the dc steady-state current voltage (I-V) characteristic of this junction. Note the following structural features and variables:

p-type bulk:
Position variable: x_p.
Hole current density at $x_p = 0$: $J_{pp}(0)$.
Electron current density at $x_p = 0$: $J_{np}(0)$.
Length: $W_P \ll L_n$.
Termination: Ohmic contact with infinite surface recombination velocity at $x_p = W_p$.

Transition region:
Position variable: x_t.
Length: W_t.
Net generation rate: $(G_{th} - R)_t$.

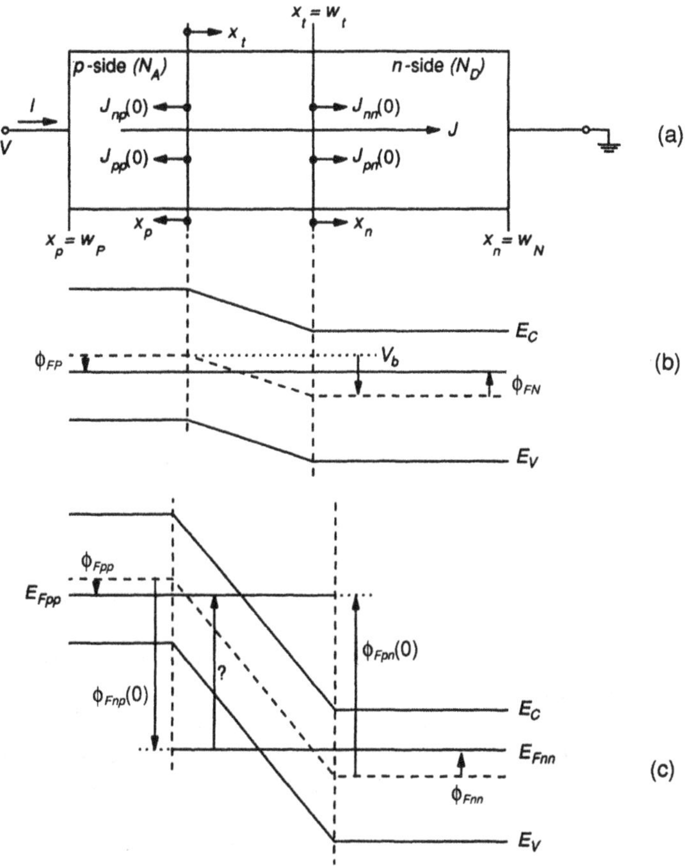

Figure P1.8

n-type bulk:
Position variable: x_n.
Hole current density at $x_n = 0$: $J_{pn}(0)$.
Electron current density at $x_n = 0$: $J_{nn}(0)$.
Length: $W_N \gg L_p$.
Doping concentration: $N = N_D$.
Condition of operation: Low-level injection.

(a) Since the structure operates under dc steady-state conditions, (1.27) implies a position-independent total current density J, which enables you to select any point to add up the electron and hole current densities to obtain J. For example, selecting $x_n = 0$ as the summing point, you can write $J = J_{pn}(0) + J_{nn}(0)$. Prove that this equation can be rewritten as

$$J = J_{pn}(0) - J_{np}(0) + J_{gr} \qquad \text{(P1.8.1)}$$

where

$$J_{gr} \equiv q\int_0^{W_t}(R - G_{th})_t dx_t$$

Select $x_p = 0$ as the summing point and repeat the derivation. Compare the results.

(b) Now go through appropriate steps to turn (P1.8.1) into

$$J = qD_n\frac{n'_p(0)}{W_P} + qD_p\frac{p'_n(0)}{L_p} + J_{gr} \quad \text{(P1.8.2)}$$

where $n'_p(0)$ and $p'_n(0)$ are the minority excess concentrations at $x_p = 0$ and $x_n = 0$, respectively.

(c) Since the diffusion constants and diffusion lengths are independent of the bias V applied, the V dependence of J should arise from $n'_p(0)$, $p'_n(0)$, and J_{gr}. You are now requested to derive the expressions relating the former two to V. To this end, first write an expression for the built-in voltage V_b in terms of N_A and N_D with the aid of the equilibrium energy-band diagram shown in Figure P1.8(b).

When a voltage V is externally applied, the junction enters nonequilibrium because a current I starts flowing. Generally, we expect some fraction of V to drop along the two bulk regions while the rest of it is superimposed on V_b. An expression for the voltage drop along a bulk region can be obtained by integrating (1.96) over the length of the region. But, if J is rather weak and N is large, this voltage will be very small. For this reason, the conventional modeling of a *pn* junction is based on the assumption of negligible bulk voltages. Naturally, therefore, V is assumed to appear across the transition region, and thus to modify the built-in voltage as depicted in Figure P1.8(c) for the case of $V < 0$. Notice the bulk majority-carrier quasi-Fermi levels E_{Fpp} and E_{Fnn} and the corresponding quasi-Fermi potentials ϕ_{Fpp} and ϕ_{Fnn}. Recalling the low-level injection condition of these bulk regions, find expressions for ϕ_{Fpp} and ϕ_{Fnn} in terms of the doping concentrations. Also find the potential difference between E_{Fpp} and E_{Fnn} as indicated by the arrow marked "?."

Assuming that the two quasi-Fermi levels do not vary significantly with position along the transition region, you can define the minority-carrier quasi-Fermi potentials $\phi_{Fpn}(0)$ and $\phi_{Fnp}(0)$ at the boundaries $x_n = 0$ and $x_p = 0$, respectively, simply by extending the levels E_{Fpp} and E_{Fnn} all the way through the transition region. This will allow you to express $p'_n(0)$ and $n'_p(0)$ as functions of V. Derive these functions, and use them in (P1.8.2) to obtain

$$J = qn_i^2\left(\frac{D_n}{W_P N_A} + \frac{D_p}{L_p N_D}\right)\left[\exp\left(\frac{q}{kT}V\right) - 1\right] + J_{gr}$$

Except for J_{gr}, this expression is traditionally known as the *diode equation*. Notice that J_{gr} is associated with the generation and recombination processes inside the transition region. The rest of the right-hand side of this equation is due to the two minority-carrier diffusion currents flowing at the boundaries of the two bulk regions.

REFERENCES

[1] Pierret, Robert F., *Modular Series on Solid State Devices, Volume I: Semiconductor Fundamentals*, 2nd ed., Reading, MA: Addison-Wesley Publishing Company, 1989.

[2] Pierret, Robert F., *Modular Series on Solid State Devices, Volume VI: Advanced Semiconductor Fundamentals,* Reading, MA: Addison-Wesley Publishing Company, 1987.

[3] Warner, Jr., R. M., and B. L. Grung, *Semiconductor-Device Electronics,* Philadelphia: Holt, Rinehart and Winston, 1991, Chaps. 1 and 2.

[4] Shur, Michael, *Physics of Semiconductor Devices,* Englewood Cliffs, NJ: Prentice Hall, 1990, Chap. 1.

[5] Ferendeci, Altan, *Physical Foundations of Solid State and Electron Devices,* New York: McGraw Hill, 1991, Chaps. 1–6.

[6] Sze, S. M. *Physics of Semiconductor Devices,* New York: John Wiley & Sons, 1981, Chap. 1.

[7] Jaeger, R. C., *Modular Series on Solid State Devices, Volume V: Introduction to Microelectronic Fabrication,* Reading, MA: Addison-Wesley Publishing Company, 1989.

[8] Hall, R. N., "Germanium Rectifier Characteristics," *Phys. Rev.*, Vol. 83, 1951, p. 228.

[9] Shockley, W., and W. T. Read, Jr., "Statistics of the Recombination of Holes and Electrons," *Phys. Rev.*, Vol. 87, 1952, p. 835.

[10] Stevenson, D. T., and R. J. Keyes, "Measurement of Carrier Lifetime in Germanium and Silicon," *J. Appl. Phys.*, Vol. 26, 1955, p. 190.

[11] Gartner, W. W., "Spectral Distribution of the Photomagnetic Electric Effect," *Phys. Rev.*, Vol. 105, 1957, p. 823.

[12] Heiman, F. P., "On the Determination of Minority Carrier Lifetime from the Transient Response of an MOS Capacitor," *IEEE Trans. on Electron Devices,* Vol. ED-14, 1967, p. 781.

[13] Pierret, Robert F., "A Linear-Sweep MOS-C Technique for Determining Minority Carrier Lifetimes," *IEEE Trans. on Electron Devices,* Vol. ED-19, 1972, p. 869.

Chapter 2
Bipolar Junction Transistors

2.1 BJTs IN THERMAL EQUILIBRIUM

The structure of a bipolar junction transistor (BJT) essentially comprises two very closely spaced *pn* junctions called *emitter-base (EB) junction* and *collector-base (CB) junction.* As shown in Figure 2.1(a), these junctions divide the structure into three regions: a heavily doped *n*-type *emitter,* a moderately doped *p*-type *base,* and finally, a *collector,* which consists of a heavily doped *n*-type *substrate* and a lightly doped *n*-type *active collector.* The basic transistor action is confined to the shaded area, which includes the emitter, base, and active collector regions. The substrate provides a mechanical support and also a low-resistance path between the active collector region and the collector contact.

The structure shown in Figure 2.1(a) is typical of a discrete BJT. In integrated circuits, all BJTs must share a common substrate for technological reasons. To provide for electrical isolation between the collectors of different transistors on the same chip, the structure is modified as shown in Figure 2.1(b). Here, each collector is surrounded by an isolation frame made of SiO_2 or *p*-type diffusion on top of a *p*-type substrate. Keeping the substrate reverse-biased prevents carrier injection between adjacent collectors, thus effecting isolation. Also included in the integrated structure is a highly doped *n*-type *buried layer,* which provides a low-resistance conduction path between the active collector region and the top-surface collector contact. More detailed structural descriptions of discrete and integrated BJTs and in-depth information on the fabrication technology can be found in Ghandi [1].

The spatial variation of the net doping concentration in the active transistor region is essentially one dimensional along *x,* as previously shown in Figure 1.15. To facilitate modeling, however, we will initially assume all three regions to be

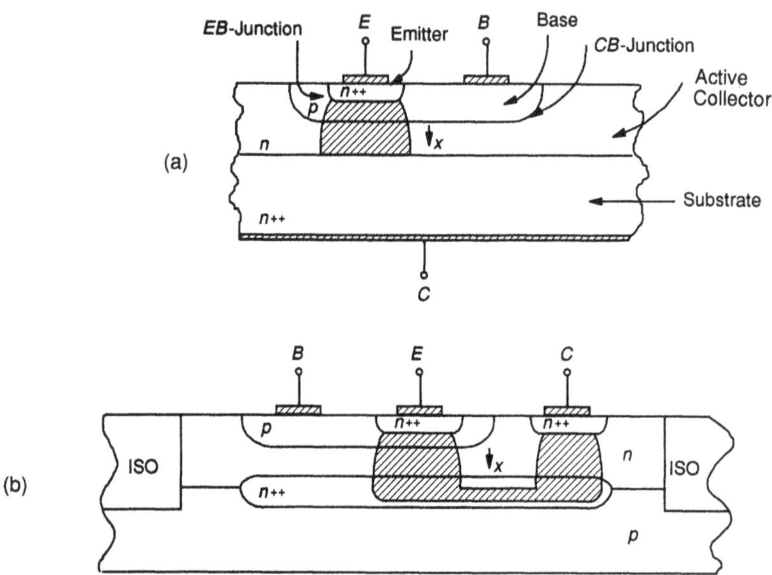

Figure 2.1 Cross-sectional schematics of BJT structures: (a) discrete BJT and (b) integrated BJT.

doped uniformly to $N = N_E$ in the emitter, $|N| = N_B$ in the base, and $N = N_C$ in the collector, as shown in Figure 2.2(a). This assumption implies $\mathscr{E} = 0$ in the emitter, base, and collector bulk regions. As already discussed in Section 1.2.3, the neighborhoods of the two junctions are space-charged and, hence, not field free. These are the so-called *transition regions.* As shown in Figure 2.2(b), the field in these two regions must be directed from the *n* side of the junction to the *p* side in order to balance the diffusion motion of the carriers with an opposing drift motion, so that current conduction is prevented as required in a thermal equilibrium state. This is why the field is positive in the EB transition region and negative in the CB transition region. In both cases, the rising edge of the field profile is located on the *n* side of the transition region, whereas the falling edge is confined to the *p* side. Poisson's equation therefore implies a positive charge on the *n* side and a negative charge on the *p* side, as depicted in Figure 2.2(c). Note that the area bounded by the $\rho = \rho(x)$ curve on the *n* side of a junction equals its counterpart on the *p* side. This can be formally proved by integrating Poisson's equation over position between the two boundaries of the transition region and, then, substituting zero for the two boundary values of the field. The resulting equation

$$\int_{n\ \text{side}} \rho dx = -\int_{p\ \text{side}} \rho dx \tag{2.1}$$

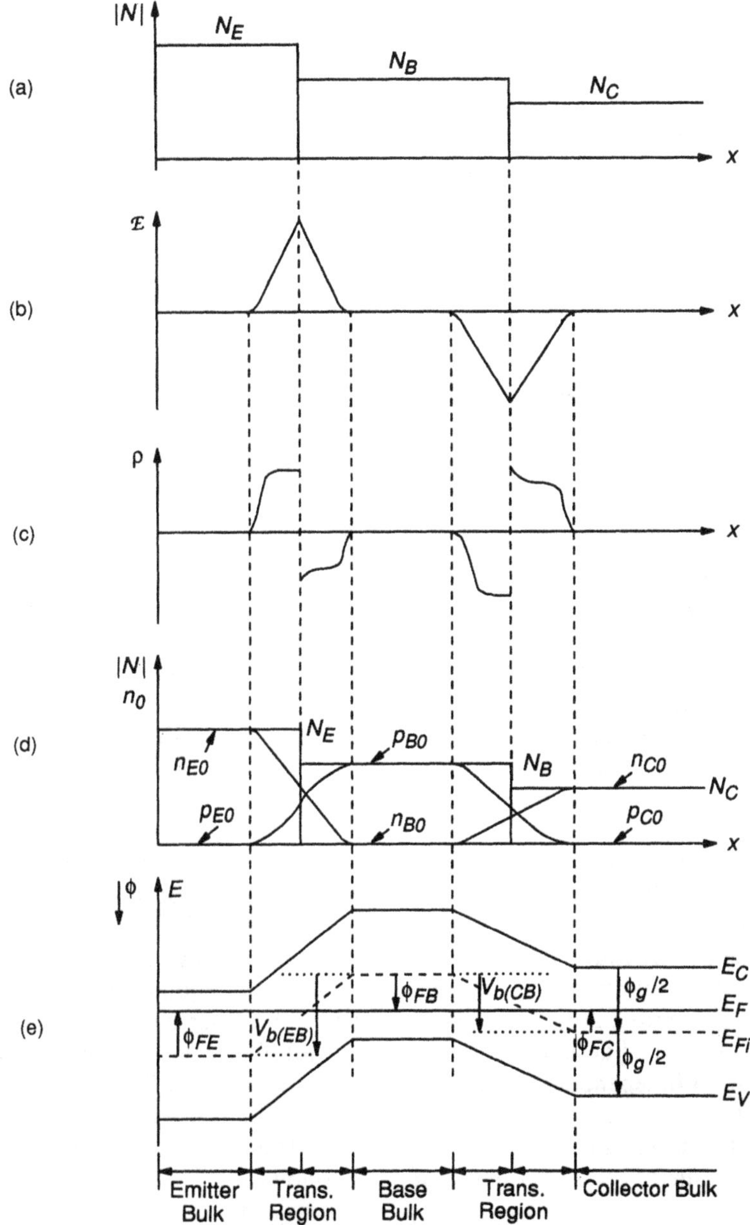

Figure 2.2 Features of a BJT in equilibrium: (a) idealized doping concentration profile, (b) electric-field profile, (c) volumetric charge density profile, (d) carrier concentration profiles, and (e) energy-band diagram.

shows that the positive charge per unit area (perpendicular to x) on the n side and the negative charge per unit area on the p side are of the same magnitude. Therefore, the charge distribution in a transition region has a dipole character. The way in which this dipole charge develops is illustrated in Figure 2.2(d) in terms of carrier concentration profiles and the doping profile. The bulk carrier concentrations, denoted in the figure by n_{Eo} and p_{Eo} for the emitter bulk, by p_{Bo} and n_{Bo} for the base bulk, and by n_{Co} and p_{Co} for the collector bulk, are determined from (1.52) through (1.55) as follows:

$$\textit{Emitter bulk: } n_{Eo} = N_E \quad \text{and} \quad p_{Eo} = \frac{n_i^2}{N_E} \tag{2.2}$$

$$\textit{Base bulk: } p_{Bo} = N_B \quad \text{and} \quad n_{Bo} = \frac{n_i^2}{N_B} \tag{2.3}$$

$$\textit{Collector bulk: } n_{Co} = N_C \quad \text{and} \quad p_{Co} = \frac{n_i^2}{N_C} \tag{2.4}$$

In transition regions, however, the carrier concentrations deviate from their respective bulk values. On the n side of the EB transition region, for example, the electron concentration falls below N_E, whereas the hole concentration rises above n_i^2/N_E. The net charge density, $\rho = q(p_o - n_o + N_E)$, which is zero in the neighboring emitter bulk, thus becomes positive. On the p side of the same transition region, the electron concentration exceeds n_i^2/N_B, whereas the hole concentration falls below N_B. This creates a negative $\rho = q(p_o - n_o - N_B)$ on that side. This is how a dipole charge develops in the transition region of a *pn* junction.

The final topic we discuss in this section is the equilibrium energy-band diagram of a BJT. Conventional BJTs are essentially homogeneous devices. In practice, however, the bandgap is slightly narrower in the emitter in comparison with the rest of the device as a result of the heavier doping of the emitter. Initially, we will ignore this bandgap-narrowing (BGN) effect, and will assume a perfectly homogeneous structure; the effect of BGN will be introduced in Section 2.2.6. Based on this assumption, the energy-band diagram in equilibrium can be plotted as shown in Figure 2.2(e). We start constructing the diagram simply by drawing a horizontal line to be designated as the Fermi level E_F because, according to (1.46), E_F is position independent. Next, we can mark the intrinsic Fermi level E_{Fi} in the three bulk regions because the difference between E_F and E_{Fi} is readily calculable from (1.41) and (1.42) and (2.2) through (2.4). The result, as converted by (1.43) into the so-called *bulk Fermi potential,* is as follows:

$$\textit{Emitter bulk: } \phi_{FE} \equiv -\frac{E_F - E_{Fi}}{q}\bigg|_{\text{emitter}} = -\frac{kT}{q}\ln\frac{N_E}{n_i} \tag{2.5}$$

$$\text{Base bulk: } \phi_{FB} \equiv -\left.\frac{E_F - E_{Fi}}{q}\right|_{\text{base}} = \frac{kT}{q}\ln\frac{N_B}{n_i} \tag{2.6}$$

$$\text{Collector bulk: } \phi_{FC} \equiv -\left.\frac{E_F - E_{Fi}}{q}\right|_{\text{collector}} = -\frac{kT}{q}\ln\frac{N_C}{n_i} \tag{2.7}$$

These bulk Fermi potentials and the intrinsic Fermi levels they define are also marked in Figure 2.2(e). Since E_{Fi} lies in the middle of the bandgap, the lowest conduction band level E_C can now be plotted one-half bandgap ($E_g/2$) above E_{Fi}. Similarly, the highest valence band level E_V is plotted one-half bandgap below E_{Fi}. Note that the potential difference corresponding to $E_g/2$ is denoted by $\phi_g/2$ in Figure 2.2(e). Also note the band bending in the two *pn* junction transition regions. In both, the bending indicates an electric field directed form the *n* side of the junction to the *p* side, as one can deduce by applying (1.31) to the diagram.

Total band bending across a transition region corresponds to the electrostatic potential difference between the two bulk regions neighboring the transition region. This potential difference is called the *built-in voltage* or *barrier potential* of the *pn* junction. In Figure 2.2(e), $V_{b(EB)}$ and $V_{b(CB)}$ denote the built-in potentials of the EB and CB junctions, respectively. Two simple loop equations written from the energy-band diagram yield $V_{b(EB)} = \phi_{FB} - \phi_{FE}$ and $V_{b(CB)} = \phi_{FB} - \phi_{FC}$. Substituting (2.5) through (2.7) into these equations, we obtain

$$V_{b(EB)} = \frac{kT}{q}\ln\frac{N_E N_B}{n_i^2} \tag{2.8}$$

$$V_{b(CB)} = \frac{kT}{q}\ln\frac{N_C N_B}{n_i^2} \tag{2.9}$$

according to which the built-in voltages are determined solely by the doping concentrations and temperature. Finally, note from (1.6) that a built-in voltage can also be expressed by integrating $\mathscr{E}$ over the transition region as

$$V_b = \pm \int_{\text{trans. region}} \mathscr{E} dx \tag{2.10}$$

Therefore, the area bounded by the field profile of a transition region directly yields the built-in voltage.

2.2 BJTs UNDER BIAS

Using certain simplifying assumptions we analyze an electrically biased BJT with a view to developing a basic low- and medium-bias model relating port variables to structural parameters. An assumption of dc steady-state conditions with low-level injection and negligible voltage drop in all three bulk regions will be maintained throughout the entire analysis. Initially, we will also impose an assumption of uniform doping-concentration profiles but, later in the analysis, we will remove it. The model development will be completed by incorporating an influential secondary effect called the *Early effect.* For supplementary reading on the subject matter of this section, the reader is referred to Roulston [2], Warner and Grung [3], and Antognetti and Massobrio [4].

2.2.1 Fundamentals of the Nonequilibrium Analysis

The BJT, being a three-terminal device, has two independent ports. When these ports are driven by external sources, the device enters a nonequilibrium state because it starts exchanging energy (via its terminals) by way of current flow. The main objective of BJT modeling in nonequilibrium is to find a relationship between these port variables and structural parameters.

We start with Figure 2.3, where the active region of a BJT is shown together with the reference directions selected for all port variables and relevant internal variables. We assume a dc steady state throughout the structure and low-level injection conditions in all three bulk regions. Under these circumstances, the bulk regions remain quasineutral in nonequilibrium, as discussed in Section 1.3.3. According to (1.6) and (1.96) or (1.99), therefore, the voltage across each bulk is described by

$$|\Delta\psi| = \frac{L}{q\mu_{n,p}}\frac{|J|}{|N|} \tag{2.11}$$

where $|J|$, L, $|N|$, and $\mu_{n,p}$ are the general variables of total current density, length, doping concentration, and majority-carrier mobility, respectively. Since bulk regions are rather short, doping concentrations are not too light, and current densities under low-level injection conditions are not too high, we expect a negligible voltage drop across each bulk. Accordingly, the port voltages V_{EB} and V_{CB} should appear directly across the EB and CB transition regions, respectively. As a result, the barrier potentials are modified as depicted in Figure 2.4(a). We will now further develop the nonequilibrium energy-band diagram before returning to Figure 2.3.

According to (1.65) and (1.66), the majority-carrier concentration in a bulk practically remains invariant as the conditions change from equilibrium to nonequilibrium under low-level injection. As indicated by a comparison between (1.44) and (1.105), or between (1.45) and (1.106), this invariance of the majority-carrier concentration implies an equality of the equilibrium Fermi potential and the majority-carrier quasi-Fermi potential. Denoting the latter by ϕ_{FnE} for the emitter

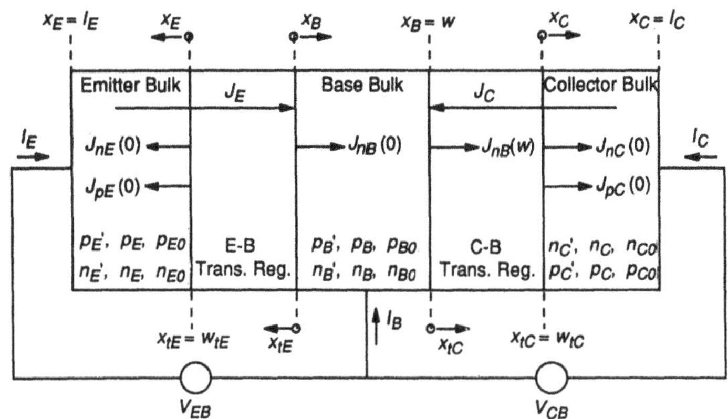

Figure 2.3 The notation and reference convention used in the nonequilibrium analysis of BJTs.

bulk, by ϕ_{FpB} for the base bulk, and by ϕ_{FnC} for the collector bulk, we therefore can write $\phi_{FnE} = \phi_{FE}$, $\phi_{FpB} = \phi_{FB}$ and $\phi_{FnC} = \phi_{FC}$. Relying on these equations we can add to the energy-band diagram the bulk majority-carrier quasi-Fermi levels E_{FnE}, E_{FpB}, and E_{FnC}, as shown in Figure 2.4(b), because ϕ_{FE}, ϕ_{FB}, and ϕ_{FC} are already known from (2.5) through (2.7).

As the next step in the construction of the nonequilibrium energy-band diagram, we extend the bulk majority-carrier quasi-Fermi levels into the neighboring transition regions, as shown in Figure 2.4(c). This is justifiable on the basis of the assumption that the low-level injection conditions prevailing in the bulk regions keep the magnitudes of J_p and J_n small everywhere in the BJT including the transition regions. For small current densities, the alternative current density equations (1.107) and (1.108) indicate a weak quasi-Fermi level gradient, which indeed justifies a straight extension of these levels into these relatively narrow transition regions.

The potential difference between the majority-carrier quasi-Fermi levels of two neighboring bulk regions is denoted in Figure 2.4(c) by V_1 for the EB junction and by V_2 for the CB junction. A loop equation involving the barrier potentials and the majority-carrier quasi-Fermi potentials of the neighboring bulk regions yields $V_1 = V_{EB}$ and $V_2 = V_{CB}$. Therefore, the potential difference between the two neighboring bulk majority-carrier quasi-Fermi levels is determined directly by the bias voltage applied to the related junction. Soon, we will make good use of these energy-band properties. Let us now return to Figure 2.3.

Between $x_E = l_E$ and $x_B = 0$, the total current density J_E must be position independent because there is no intermediate terminal to sink or source current and (1.27) is unconditionally valid in a dc steady state. Therefore, the sum of

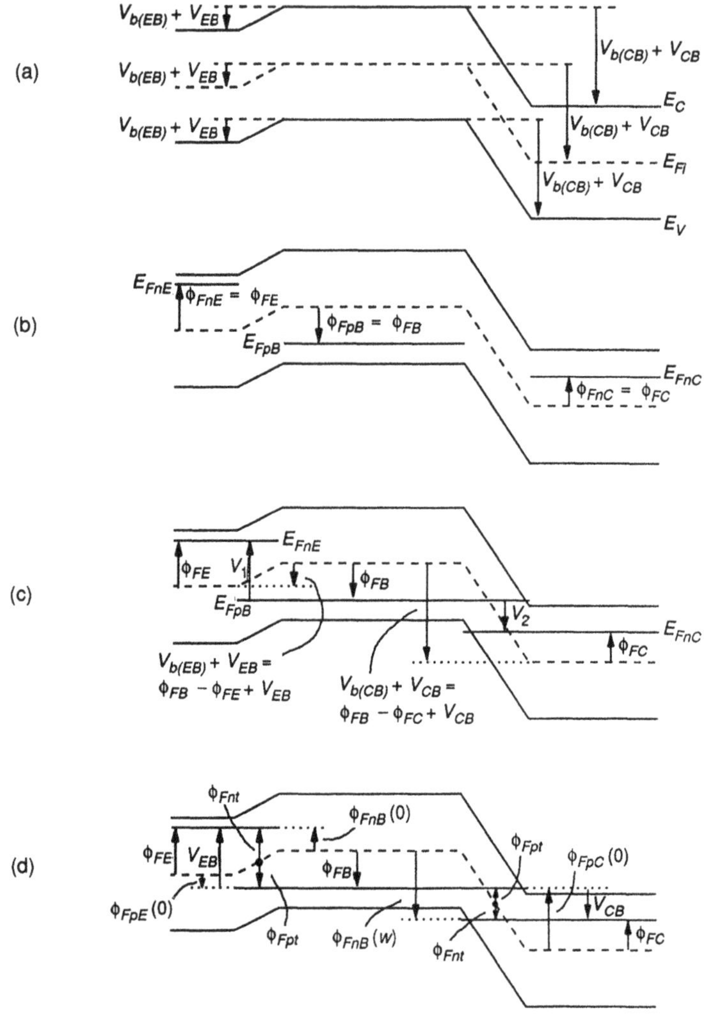

Figure 2.4 Stages of energy-band diagram construction for a BJT in nonequilibrium: (a) modified barrier potentials, (b) adding the bulk majority-carrier quasi-Ferm: levels, (c) extending the bulk majority-carrier quasi-Fermi levels, and (d) energy-band diagram.

electron and hole current densities at any point in this region should give the density J_E of the emitter terminal current. Selecting $x_E = 0$ as the summing point, we can write

$$J_E = -J_{pE}(0) - J_{nE}(0) \tag{2.12}$$

The fact that point $x_E = 0$ belongs to a bulk region operating under low-level injection conditions enables us to ignore the drift component of the minority-carrier current $J_{pE}(0)$, as discussed in Section 1.3.3. This tremendously simplifies the formulation of $J_{pE}(0)$. On the other hand, $J_{nE}(0)$ is a majority-carrier current density, which is not amenable to such a simplification. Fortunately, we can express it in terms of $J_{nB}(0)$, which, being a minority-carrier current, can also be easily formulated. For this purpose, let us integrate the current continuity equation (1.26) over the EB transition region assuming $g = 0$. The result is

$$J_{nE}(0) = -J_{nB}(0) - J_{grE} \tag{2.13}$$

where

$$J_{grE} \equiv q\int_0^{W_{tE}} (G_{th} - R)\, dx_{tE} \tag{2.14}$$

is called EB transition region current density. Substituting (2.13) into (2.12) results in

$$J_E = -J_{pE}(0) + J_{nB}(0) + J_{grE} \tag{2.15}$$

Repeating this derivation for the collector current density J_C by selecting $x_C = 0$ as the summing point, we obtain

$$J_C = -J_{pC}(0) - J_{nB}(W) + J_{grC} \tag{2.16}$$

where

$$J_{grC} \equiv q\int_0^{W_{tC}} (G_{th} - R)\, dx_{tC} \tag{2.17}$$

is called CB transition region current density.

In Sections 2.2.2 and 2.2.3, we will relate all components of J_E and J_C to bias voltages and structural parameters. The results will be used in Section 2.2.4 to derive the characteristic equations of a BJT under bias.

2.2.2 Bulk Current Components

Emitter Bulk: J_{pE}

Ignoring the drift component, $J_{pE}(0)$ can be expressed as

$$J_{pE}(0) \cong -qD_{pE}\left.\frac{dp_E}{dx_E}\right|_{x_E=0} = -qD_{pE}\left.\frac{dp'_E}{dx_E}\right|_{x_E=0} \tag{2.18}$$

For determining the minority-carrier excess concentration profile $p'_E(x_E)$, we assume that $l_E \ll L_{pE}$, where L_{pE} is the hole diffusion length in the emitter, and we also assume $p'_E(l_E) = 0$ because an ohmic contact is placed at $x_E = l_E$. The appropriate solution is obtained from (1.90) as

$$p'_E = \left(1 - \frac{x_E}{l_E}\right)p'_E(0) = \left(1 - \frac{x_E}{l_E}\right)[p_E(0) - p_{Eo}] \tag{2.19}$$

where $p_E(0)$ is described by (1.106) as

$$p_E(0) = n_i \exp\left[\frac{q}{kT}\phi_{FpE}(0)\right] \tag{2.20}$$

and the hole quasi-Fermi potential at the emitter boundary, $\phi_{FpE}(0)$, is marked on the energy-band diagram of Figure 2.4(d). A simple loop equation involving $\phi_{FpE}(0)$, ϕ_{FE}, and V_{EB} yields $\phi_{FpE}(0) = \phi_{FE} - V_{EB}$. Substituting this into (2.20) and recognizing the equation

$$n_i \exp\left(\frac{q}{kT}\phi_{FE}\right) = p_{Eo} = \frac{n_i^2}{N_E}$$

we can write

$$p_E(0) = p_{Eo} \exp\left(-\frac{q}{kT}V_{EB}\right) = \frac{n_i^2}{N_E} \exp\left(-\frac{q}{kT}V_{EB}\right) \tag{2.21}$$

From (2.21), (2.19), and (2.18), we finally obtain

$$J_{pE}(0) = \frac{qD_{pE}n_i^2}{N_E l_E}\left[\exp\left(-\frac{q}{kT}V_{EB}\right) - 1\right] \tag{2.22}$$

Base Bulk: J_{nB}

The current components we will examine next are $J_{nB}(0)$ and $J_{nB}(W)$. These, being the densities of the minority-carrier currents flowing at the two boundaries of the *p*-type base bulk, can be expressed as

$$J_{nB}(0) = qD_{nB}\left.\frac{dn'_B}{dx_B}\right|_{x_B=0} \tag{2.23}$$

$$J_{nB}(W) = qD_{nB}\left.\frac{dn'_B}{dx_B}\right|_{x_B=W} \tag{2.24}$$

For a modern BJT with a narrow base, one can assume $W \ll L_{nB}$ for which $n'_B(x_B)$ can be obtained from (1.94) as

$$n'_B = \left(1 - \frac{x_B}{W}\right)[n_B(0) - n_{Bo}] + \frac{x_B}{W}[n_B(W) - n_{Bo}] \tag{2.25}$$

where $n_B(0)$ and $n_B(W)$, as derived from the energy-band diagram of Figure 2.4(d), are as follows:

$$n_B(0) = \frac{n_i^2}{N_B} \exp\left(-\frac{q}{kT}V_{EB}\right) \tag{2.26}$$

$$n_B(W) = \frac{n_i^2}{N_B} \exp\left(-\frac{q}{kT}V_{CB}\right) \tag{2.27}$$

From (2.23) through (2.27), we finally obtain

$$J_{nB}(0) = J_{nB}(W) = -\frac{qD_{nB}n_i^2}{N_BW}\left[\exp\left(-\frac{q}{kT}V_{EB}\right) - \exp\left(-\frac{q}{kT}V_{CB}\right)\right] \tag{2.28}$$

Collector Bulk: J_{pC}

The final minority-carrier bulk current we need to examine is $J_{pC}(0)$, which is described by

$$J_{pC}(0) = -qD_{pC}\frac{dp'_C}{dx_C}\bigg|_{x_C=0} \tag{2.29}$$

Since l_C is assumed to be much larger than the minority-carrier diffusion length in a typical collector, $p'_C(x_C)$ is described by (1.89) as follows:

$$p'_C = p'_C(0) \exp\left(-\frac{x_C}{L_{pC}}\right) = [p_C(0) - p_{Co}] \exp\left(-\frac{x_C}{L_{pC}}\right) \tag{2.30}$$

where $L_{pC} \equiv \sqrt{D_{pC}\tau_{pC}}$ is the minority-carrier diffusion length in the collector bulk. For $p_C(0)$ we refer to the energy-band diagram of Figure 2.4(d):

$$p_C(0) = \frac{n_i^2}{N_C} \exp\left(-\frac{q}{kT}V_{CB}\right) \tag{2.31}$$

From (2.29) through (2.31), we finally obtain

$$J_{pC}(0) = \frac{qD_{pC}n_i^2}{N_CL_{pC}}\left[\exp\left(-\frac{q}{kT}V_{CB}\right) - 1\right] \tag{2.32}$$

2.2.3 Transition Region Current Components

Net Generation Rate in a Transition Region

As indicated by (2.14) and (2.17), the transition region current components J_{grE} and J_{grC} have similar definitions. In both, the net generation rate must be integrated over the transition region involved. For $G_{th} - R$, we must use (1.73) because the simpler form of (1.78) or (1.80) is not valid in a transition region.

The *np* product in (1.73) can be expressed in terms of the quasi-Fermi potentials ϕ_{Fnt} and ϕ_{Fpt} of the transition region as

$$n_t p_t = \left[n_i \exp\left(-\frac{q}{kT}\phi_{Fnt}\right)\right]\left[n_i \exp\left(\frac{q}{kT}\phi_{Fpt}\right)\right] = n_i^2 \exp\left[\frac{q}{kT}(\phi_{Fpt} - \phi_{Fnt})\right]$$

Note that no subscript is used in this equation to indicate a particular transition region because the expression is valid for both transition regions. As indicated in the energy-band diagram of Figure 2.4(d), the difference $\phi_{Fnt} - \phi_{Fpt}$ equals the external bias V of the junction, that is

$$n_t p_t = n_i^2 \exp\left(-\frac{q}{kT}V\right) \tag{2.33}$$

where V represents V_{EB} or V_{CB}. Using this result together with $\sigma \equiv \sigma_n = \sigma_p$ in (1.73), we arrive at

$$G_{th} - R = \frac{v_{th}\sigma n_i}{2}\left[1 - \exp\left(-\frac{q}{kT}V\right)\right]\int_{-E_g/2}^{E_g/2} \frac{D_t d(E - E_{Fi})}{\dfrac{n_t + p_t}{2n_i} + \cosh\dfrac{E - E_{Fi}}{kT}} \tag{2.34}$$

It is obvious from (2.34) that $G_{th} - R$ is positive for $V > 0$, which defines a *reverse bias,* and negative for $V < 0$, which defines a *forward bias.* In other words, the dominant process in the transition region is *generation* for a reverse bias and *recombination* for a forward bias.

The Case of Reverse Bias

Equation (2.34) can be further simplified if we specify the type of bias involved. First consider a reverse-biased junction. The energy-band diagram of the CB junction in Figure 2.4(d) is drawn for this kind of bias. It is obvious that, in a large portion of the transition region, the conditions $\phi_{Fnt} > 0$ and $\phi_{Fpt} < 0$ coexist. This implies $n_t < n_i$ and $p_t < n_i$, which enables us to ignore $(n_t + p_t)/2n_i$ in the denominator of the integrand of (2.34). Using this simplified form of (2.34) in (2.14) or in (2.17), we obtain

$$J_{gr} = \frac{qn_iW_t}{2\tau_{or}}\left[1 - \exp\left(-\frac{q}{kT}V\right)\right], \qquad \text{for } V > 0 \tag{2.35}$$

where

$$\tau_{or} \equiv \frac{1}{v_{th}\sigma\int_{-E_g/2}^{E_g/2}\left[\cosh\left(\frac{E - E_{Fi}}{kT}\right)\right]^{-1} D_t\, d(E - E_{Fi})} \qquad \text{for } V > 0 \tag{2.36}$$

is the effective lifetime in a reverse-biased transition region. The term $\{\cosh[(E - E_{Fi})/kT]\}^{-1}$ of (2.36) equals unity for the traps located in the middle of the bandgap ($E - E_{Fi} = 0$). Moving away from the midgap in either direction, this term decreases exponentially, and thus forces the integrand to vanish. Therefore, only the traps lying inside a very narrow midgap energy range can participate in the generation process.

The Case of Forward Bias

In the case of a forward-biased junction, we expect $(n_t + p_t)/2n_i \gg 1$ because, as indicated by the EB transition region energy-band diagram in Figure 2.4(d), a forward bias causes $\phi_{Fpt} > 0$ and $\phi_{Fnt} > 0$. Now consider the denominator of the integrand in (2.34). For those traps whose energies satisfy $|E - E_{Fi}| < kT\ln[(n_t + p_t)/n_i]$, the term $\cosh[(E - E_{Fi})/kT]$ is negligible. For those with a higher $|E - E_{Fi}|$, the term $(n_t + p_t)/2n_i$ is negligible; but, since $\cosh[(E - E_{Fi})/kT]$ is also very large for such traps, the integrand becomes large enough to reduce $G_{th} - R$ to a negligible level. Therefore, the traps that can effectively contribute to the recombination process are those located at energies within the midgap range $\pm kT\ln[(n_t + p_t)/2n_i]$. This enables us to rewrite (2.34) as

$$G_{th} - R = \frac{v_{th}\sigma n_i^2}{n_t + p_t}\left[1 - \exp\left(-\frac{q}{kT}V\right)\right]\int_{-E_T}^{E_T} D_t\, d(E - E_{Fi}), \qquad \text{for } V < 0 \tag{2.37}$$

where $E_T \equiv kT\ln[(n_t + p_t)/n_i]$. The net generation rate in a forward-biased transition region appears to be a rather complex function of position due to its dependence on n_t and p_t. For this reason, its integration over position as required by (2.14) or (2.17) does not yield an analytical result for the current. To overcome this problem, $G_{th} - R$ in the case of forward bias is assumed constant throughout the transition region at its maximum value. According to (2.37), $G_{th} - R$ is a maximum at the point where $n_t + p_t$ is minimized, which is the point where $n_t = p_t$ because the product $n_t p_t$ is position independent as implied by (2.33). Substituting $n_t = p_t$ into the latter, using in (2.37) the minimum value of $n_t + p_t$ thus found, and integrating in accordance with (2.14) or (2.17), we arrive at

$$J_{gr} \cong \frac{qn_i W_t}{2\tau_{of}} \exp\left(\frac{q}{2kT}V\right)\left[1 - \exp\left(-\frac{q}{kT}V\right)\right], \qquad \text{for } V < 0 \tag{2.38}$$

where

$$\tau_{of} \equiv \frac{1}{v_{th}\sigma \int_{-E_T}^{E_T} D_t \, d(E - E_{Fi})}, \qquad \text{for } V < 0 \tag{2.39}$$

is the effective carrier lifetime in a forward-biased transition region.

The Modeling Equations

We now complete this section by rewriting (2.35) and (2.38) with the BJT notation:

$$J_{grE} = \begin{cases} -\dfrac{qn_i W_{tE}}{2\tau_{of}} \exp\left(\dfrac{q}{2kT}V_{EB}\right)\left[\exp\left(-\dfrac{q}{kT}V_{EB}\right) - 1\right], & \text{for } V_{EB} < 0 \quad (2.40) \\[2ex] -\dfrac{qn_i W_{tE}}{2\tau_{or}}\left[\exp\left(-\dfrac{q}{kT}V_{EB}\right) - 1\right], & \text{for } V_{EB} > 0 \quad (2.41) \end{cases}$$

$$J_{grC} = \begin{cases} -\dfrac{qn_i W_{tC}}{2\tau_{of}} \exp\left(\dfrac{q}{2kT}V_{CB}\right)\left[\exp\left(-\dfrac{q}{kT}V_{CB}\right) - 1\right], & \text{for } V_{CB} < 0 \quad (2.42) \\[2ex] -\dfrac{qn_i W_{tC}}{2\tau_{or}}\left[\exp\left(-\dfrac{q}{kT}V_{CB}\right) - 1\right], & \text{for } V_{CB} > 0 \quad (2.43) \end{cases}$$

2.2.4 The Characteristic Equations of a BJT

Substituting (2.22), (2.28), (2.32) and (2.40) through (2.43) into (2.15) and (2.16), and multiplying the resulting equations by the cross-sectional area A we obtain the characteristic equations of a BJT under bias:

$$I_E = -\frac{I_S}{\alpha_F}\left[\exp\left(-\frac{q}{kT}V_{EB}\right) - 1\right] + I_S\left[\exp\left(-\frac{q}{kT}V_{CB}\right) - 1\right] \tag{2.44}$$

$$I_C = I_S\left[\exp\left(-\frac{q}{kT}V_{EB}\right) - 1\right] - \frac{I_S}{\alpha_R}\left[\exp\left(-\frac{q}{kT}V_{CB}\right) - 1\right] \tag{2.45}$$

and, since $I_B = -I_E - I_C$,

$$I_B = \frac{1-\alpha_F}{\alpha_F}I_S\left[\exp\left(-\frac{q}{kT}V_{EB}\right) - 1\Big)\right] + \frac{1-\alpha_R}{\alpha_R}I_S\left[\exp\left(-\frac{q}{kT}V_{CB}\right) - 1\Big)\right] \tag{2.46}$$

where the so-called *intercept current* I_S, *forward alpha* α_F, and *reverse alpha* α_R are defined by

$$I_S \equiv \frac{Aqn_i^2 D_{nB}}{WN_B} \tag{2.47}$$

$$\alpha_F \equiv \begin{cases} \left[1 + \dfrac{D_{pE}WN_B}{D_{nB}l_E N_E} + \dfrac{W_{tE}WN_B}{2n_i\tau_{of}D_{nB}}\exp\left(\dfrac{q}{2kT}V_{EB}\right)\right]^{-1}, & \text{for } V_{EB} < 0 \quad (2.48) \\[2ex] \left(1 + \dfrac{D_{pE}WN_B}{D_{nB}l_E N_E} + \dfrac{W_{tE}WN_B}{2n_i\tau_{or}D_{nB}}\right)^{-1}, & \text{for } V_{EB} > 0 \quad (2.49) \end{cases}$$

$$\alpha_R \equiv \begin{cases} \left[1 + \dfrac{D_{pC}WN_B}{D_{nB}L_{pC}N_C} + \dfrac{W_{tC}WN_B}{2n_i\tau_{of}D_{nB}}\exp\left(\dfrac{q}{2kT}V_{CB}\right)\right]^{-1}, & \text{for } V_{CB} < 0 \quad (2.50) \\[2ex] \left(1 + \dfrac{D_{pC}WN_B}{D_{nB}L_{pC}N_C} + \dfrac{W_{tC}WN_B}{2n_i\tau_{or}D_{nB}}\right)^{-1}, & \text{for } V_{CB} > 0 \quad (2.51) \end{cases}$$

Equations (2.44), (2.45), and (2.46), which characterize the terminal behavior of a biased BJT, are called *Ebers-Moll equations* [5]. Often these equations are expressed in the following form:

$$I_B = \frac{I_S}{\beta_F}\left[\exp\left(-\frac{q}{kT}V_{EB}\right) - 1\right] + \frac{I_S}{\beta_R}\left[\exp\left(-\frac{q}{kT}V_{CB}\right) - 1\right] \tag{2.52}$$

$$I_C = I_S\left[\exp\left(-\frac{q}{kT}V_{EB}\right) - 1\right] - \frac{1+\beta_R}{\beta_R}I_S\left[\exp\left(-\frac{q}{kT}V_{CB}\right) - 1\right] \tag{2.53}$$

and, using $I_E = -I_C - I_B$,

$$I_E = -\frac{1+\beta_F}{\beta_F} I_S\left[\exp\left(-\frac{q}{kT}V_{EB}\right)-1\right] + I_S\left[\exp\left(-\frac{q}{kT}V_{CB}\right)-1\right] \quad (2.54)$$

where the *forward beta* β_F and the *reverse beta* β_R are defined by

$$\beta_F \equiv \frac{\alpha_F}{1-\alpha_F} \quad (2.55)$$

$$\beta_R \equiv \frac{\alpha_R}{1-\alpha_R} \quad (2.56)$$

and, therefore, can be related to structural parameters as follows

$$\beta_F = \begin{cases} \left[\dfrac{D_{pE}WN_B}{D_{nB}l_EN_E} + \dfrac{W_{tE}WN_B}{2n_i\tau_{of}D_{nB}}\exp\left(\dfrac{q}{2kT}V_{EB}\right)\right]^{-1}, & \text{for } V_{EB} < 0 \quad (2.57) \\ \left(\dfrac{D_{pE}WN_B}{D_{nB}l_EN_E} + \dfrac{W_{tE}WN_B}{2n_i\tau_{or}D_{nB}}\right)^{-1}, & \text{for } V_{EB} > 0 \quad (2.58) \end{cases}$$

$$\beta_R = \begin{cases} \left[\dfrac{D_{pC}WN_B}{D_{nB}L_{pC}N_C} + \dfrac{W_{tC}WN_B}{2n_i\tau_{of}D_{nB}}\exp\left(\dfrac{q}{2kT}V_{CB}\right)\right]^{-1}, & \text{for } V_{CB} < 0 \quad (2.59) \\ \left(\dfrac{D_{pC}WN_B}{D_{nB}L_{pC}N_C} + \dfrac{W_{tC}WN_B}{2n_i\tau_{or}D_{nB}}\right)^{-1}, & \text{for } V_{CB} > 0 \quad (2.60) \end{cases}$$

2.2.5 Forward Active Operation

The Four Possible Operation Modes

The four possible bias arrangements in which a BJT can be operated are called operation modes. If both junctions are forward biased, the mode is called *saturation. Cutoff* is the mode in which both junctions are reverse biased. Forward biasing the CB junction and reverse biasing the EB junction results in the *reverse active* mode. The opposite case is called the *forward active* mode. The Ebers-Moll equations are perfectly valid in all these modes.

Being involved in both analog and digital circuit applications, the forward active mode can be regarded as the most significant of all four operation modes. We will now present a physical interpretation of the preceding section's results specifically for this mode of operation.

A Physical Interpretation of the Forward Active Mode

The BJT of Figure 2.3 is biased into the forward active mode by $V_{EB} < 0$ and $V_{CB} > 0$. The forward bias across the EB junction reduces the potential barrier, thus lowering the electric-field intensity in the transition region because the barrier is the spatial integral of the field as expressed by (2.10). In consequence, the balance between the drift and diffusion currents is upset in favor of the latter. Electrons diffuse from the emitter to the base, while holes diffuse in the opposite direction. As indicated by (2.26) and (2.21), and shown in Figure 2.5(a), this two-way flow increases electron concentration at the emitter boundary of the base and hole concentration at the base boundary of the emitter above the equilibrium levels. At the contact side of the emitter bulk, however, the high density of surface traps does not permit any excess concentration buildup, as a result of which, a negative hole concentration gradient develops in the emitter bulk. Since the emitter is assumed to be much narrower than the diffusion length of the holes, no recombination effect is observed on the hole concentration gradient; therefore, the hole concentration profile is linear there.

Turning our attention to the reverse-biased CB junction, we observe an increased barrier height, which enhances the intensity of the electric field in the transition region. This strengthens drift tendencies, causing those electrons approaching the region from the base side to be accelerated toward the collector bulk. As a result, the electron concentration at x_B = W vanishes. The enhanced field in the transition region also accelerates the holes at $x_C = 0$ toward the base, causing the hole concentration to vanish at that boundary. However, the hole concentration in the collector bulk is restored to its equilibrium level by generation within a distance of a few diffusion lengths. As observed from Figure 2.5(a), the minority-carrier electron concentration in the base has a negative gradient, which is position independent because the assumption of a very narrow base precludes any significant generation or recombination in the region.

The electron and the hole current profiles in the forward active mode are illustrated in Figure 2.5(d). The actual direction of current flow is indicated by the arrows. We start constructing these profiles by plotting the electron current in the base and the hole current in the emitter and collector. These currents flow mainly by diffusion because the carriers involved are minority carriers. Since a diffusion current is proportional to the gradient of the carrier concentration, we can construct the profiles of these currents with the aid of Figure 2.5(a). The result is depicted in Figure 2.5(b). Note that I_{nB} is much larger than I_{pE} mainly because $N_E \gg N_B$ [see (2.22) and (2.28)]. The reason for I_{pC} being much smaller than I_{nB} is the very weak gradient of p_C in comparison with the very strong gradient of n_B.

Next, we extend the electron current profile to the transition regions as shown in Figure 2.5(c). For a qualitative explanation of the position dependence of the electron current in transition regions, first consider the electrons injected by the

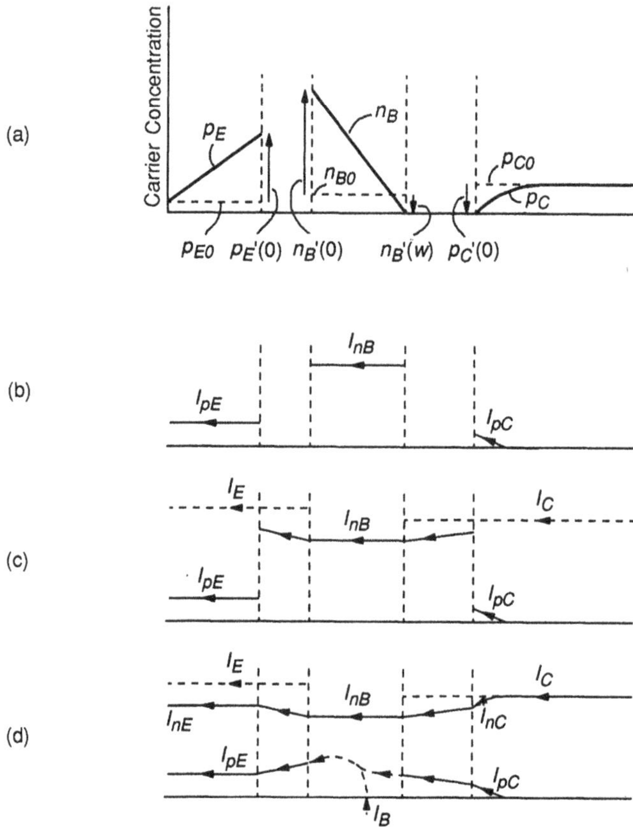

Figure 2.5 Profiles of internal variables in the forward active mode of operation: (a) minority-carrier concentration profiles, (b-d) current profiles as determined in successive stages of the analysis.

emitter into the EB transition region. Some of these electrons, while diffusing through the region, recombine with the holes injected from the base in the opposite direction. Since each recombination event annihilates one electron, the electron current decreases with position toward the base. The situation in the CB junction is different in the sense that generation, and not recombination, prevails there as a result of reverse bias. Each electron thus generated joins the electrons approaching from the base and moves on to the collector. The electron current in the CB transition region, therefore, increases toward the collector. At this point, we can also determine the total current at $x_E = 0$ and $x_C = 0$ by summing the electron and hole currents. These total currents represent the emitter and collector terminal currents, respectively, as indicated by I_E and I_C in Figure 2.5(c). As shown by the dashed lines, these total currents are position independent because the BJT is in a dc steady state. In the final step, we obtain the majority-carrier electron current profile in the emitter and collector bulk regions by subtracting the minority-carrier hole current from the position-independent I_E and I_C, respectively. The hole current profile in a transition region can be extracted similarly by subtracting the electron current profile from the relevant terminal current. With these additions, the current profiles inside the BJT take the form shown in Figure 2.5(d). We observe from this diagram that the hole current injected from base to emitter has a different value than the hole current entering the base from the collector. Obviously, the difference must be supplied by the base contact in the form of a base terminal current, I_B. The actual hole current profile in the base is very complicated as a result of the three-dimensional structure of the base region. The dashed curves representing this current in Figure 2.5(d) do not indicate the actual profile; they only indicate the direction of current flow.

Let us now summarize the ideal BJT action in the forward active mode.

1. Due to a forward bias across the EB junction, electrons are injected from the emitter into the EB transition region. The dominant component of the emitter current is created by this injection. Some of these electrons recombine in the transition region, but most reach the base and subsequently diffuse to the CB transition region. During this transportation, electrons suffer negligible recombination in the narrow base bulk. Having reached the CB transition region, these electrons are joined by those generated there, and they altogether drift into the collector bulk, and thus create the dominant component of the collector current.
2. The other component of the collector current is due to hole injection from the collector to the base. However, the reverse bias across the CB junction limits this injection to a very low level.
3. Holes are also injected from the base to the emitter through the forward-biased EB junction. However, this contribution to the emitter current is several orders of magnitude smaller than the contribution of electron injection in the opposite direction because the emitter is more heavily doped in comparison with the base.
4. The difference between the hole current flowing from base to emitter and the hole current flowing from collector to base is supplied by the base contact as the terminal base current.

Gummel Plots and the Low-Current Beta Rolloff Effect

We now examine the terminal currents as functions of V_{BE} in the forward active mode. Assuming $V_{CB} \gg (kT/q)$, the Ebers-Moll equations (2.52) and (2.53) can be reduced to

$$I_B = \frac{I_S}{\beta_F}\left[\exp\left(\frac{q}{kT}V_{BE}\right) - 1\right] - \frac{I_S}{\beta_R} \tag{2.61}$$

$$I_C = I_S\left[\exp\left(\frac{q}{kT}V_{BE}\right) - 1\right] + \frac{1 + \beta_R}{\beta_R}I_S \tag{2.62}$$

where β_F and β_R are described by (2.57) and (2.60), respectively. Figure 2.6 shows the so-called *Gummel plot,* that is, a plot of I_C and I_B versus V_{BE} on a semilogarithmic coordinate system, as predicted by (2.61) and (2.62) for a typical set of BJT parameters. Except for very low bias, I_C exhibits an exponential dependence on V_{BE}, as defined by the first term on the right-hand side of (2.62). The rate of this dependence is about 1.74 decades per 100 millivolts of variation in V_{BE}. Turning our attention to the variation of I_B with V_{BE}, we observe a negative base current for small values of bias. This is due the second term on the right-hand side of (2.61). After changing polarity for about V_{BE} = 100 mV, I_B starts increasing at a rate determined by $\exp(qV_{BE}/2kT)$, that is, 0.87 decades per 100 mV of bias variation. This rate arises from the fact that, for a relatively small V_{BE}, the second term in (2.57) (which is due to recombination in the EB transition region) is dominant. In other words, β_F is proportional to $\exp(qV_{BE}/2kT)$ for small values of V_{BE}, and therefore for small I_C, as shown in Figure 2.7. This bias-dependence

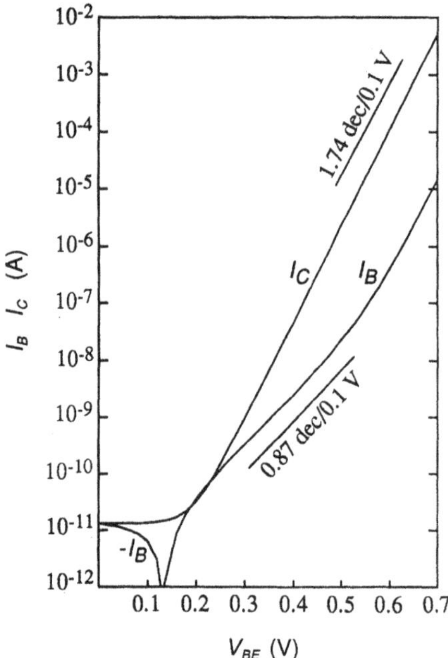

Figure 2.6 Gummel plots of a typical BJT as calculated from (2.61) and (2.62).

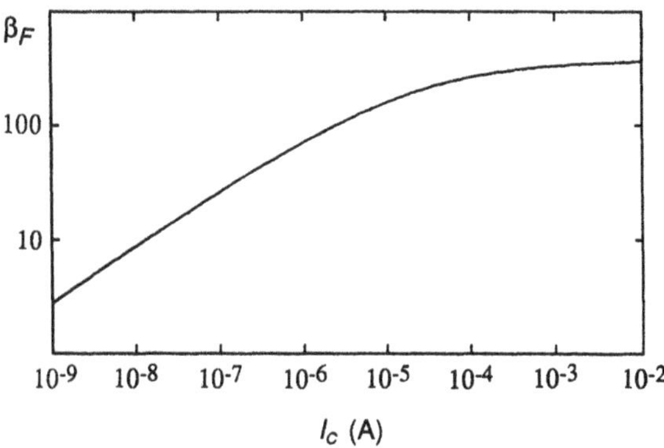

Figure 2.7 The low- and medium-current β_F as calculated from (2.57). Notice the manifestation of the "low-current beta roll-off effect" for collector current less than about 100 μA.

of β_F is known as the *low-current beta roll-off effect.* The variation of β_F with V_{BE} starts slowing down at about $V_{BE} = 400$ mV, and eventually disappears because an increasingly positive V_{BE} reduces the transition region recombination term in (2.57). The remaining term is, of course, independent of V_{BE}. For relatively higher values of V_{BE}, therefore, the variation of I_B with V_{BE} follows $\exp(qV_{BE}/kT)$ corresponding to a rate of 1.74 decades per 100 mV of bias variation, as shown in the Gummel plot of Figure 2.6.

The model we have used to develop these Gummel plots is based on an assumption of low-level injection in all bulk regions. This assumption starts losing its validity as V_{BE} exceeds the 500- to 600-mV range. The physical effects dominating the high-current operation of a BJT will be discussed in Section 2.3, where we show how the Gummel plots are modified by these effects.

2.2.6 Modeling for Nonuniform Doping Profiles

As we already know from Section 2.1, the emitter and base doping profiles of a real BJT are generally nonuniform. In this section, we will examine the effect of these nonuniform doping profiles on the Ebers-Moll equations.

Among the five components entering the terminal current equations (2.15) and (2.16), J_{grE}, J_{grC}, and $J_{pC}(0)$ are virtually uneffected by the nonuniformity of the doping profiles because the former two are associated with the transition regions, for which the analysis is only loosely related to the doping profile, and also because the collector bulk, where $J_{pC}(0)$ flows, is indeed uniformly doped. We, therefore, need only to reconsider J_{nB} and J_{pE}, hence, the base and emitter bulk regions. The nonuniform doping profiles give rise to a nonzero electric field in these regions in equilibrium, as discussed previously and as shown in Figure 1.15. Our analysis will be based on the assumption that this field profile remains virtually unchanged in nonequilibrium as a result of low-level injection in the emitter and base bulk regions. We will also retain the assumption that these regions operate under dc steady-state conditions and are free of generation or recombination except for the ohmic contact at the outer boundary of the emitter bulk. Figure 2.8 shows the reference convention adopted for position variables.

Starting with the base, we write the general current density equation for electrons as

$$J_{nB} = q\mu_{nB} n_B \mathscr{E}_B + qD_{nB}\frac{dn_B}{dx_B} \tag{2.63}$$

Rewriting this equation for equilibrium as

$$0 = q\mu_{nB} n_{Bo} \mathscr{E}_B + q\frac{kT}{q}\mu_{nB}\frac{dn_{Bo}}{dx_B}$$

and solving for $\mathscr{E}_B$, we obtain

$$\mathscr{E}_B = -\frac{kT}{q}\frac{1}{n_{Bo}}\frac{dn_{Bo}}{dx_B}$$

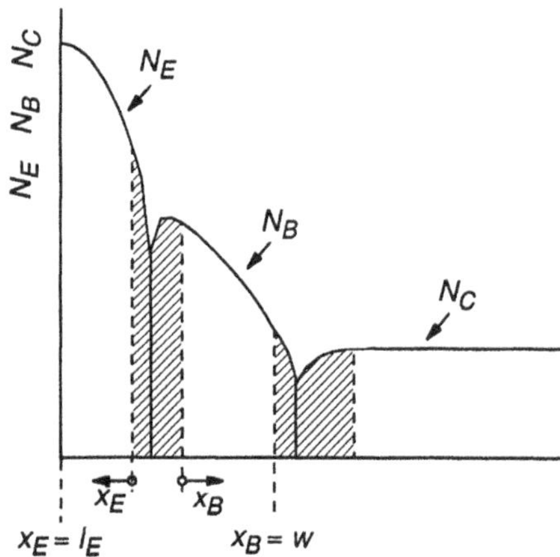

Figure 2.8 The reference convention adopted for the modeling of a BJT with nonuniform doping profiles.

Assuming that $\mathscr{E}_B$ remains invariant under nonequilibrium conditions, we can substitute this equation into (2.63) for $\mathscr{E}_B$, and thus arrive at

$$J_{nB} = qD_{nB}\left(\frac{dn_B}{dx_B} - \frac{n_B}{n_{Bo}}\frac{dn_{Bo}}{dx_B}\right) = qD_{nB}n_{Bo}\frac{d}{dx_B}\left(\frac{n_B}{n_{Bo}}\right)$$

Considering the fact that J_{nB} is position independent because the base is virtually free of generation and recombination effects, this equation can be integrated over the base as

$$J_{nB}\int_0^W \frac{dx_B}{n_{Bo}(x_B)} = qD_{nB}\left[\frac{n_B(W)}{n_{Bo}(W)} - \frac{n_B(0)}{n_{Bo}(0)}\right] \tag{2.64}$$

into which we can substitute

$$n_{Bo}(x_B) = \frac{n_i^2}{N_B(x_B)} \tag{2.65}$$

together with the boundary conditions

$$n_B(W) = n_{Bo}(W)\exp\left(-\frac{q}{kT}V_{CB}\right)$$

$$n_B(0) = n_{Bo}(0)\exp\left(-\frac{q}{kT}V_{EB}\right)$$

and thus obtain

$$J_{nB} = -\frac{qD_{nB}n_i^2}{G_b}\left[\exp\left(-\frac{q}{kT}V_{EB}\right) - \exp\left(-\frac{q}{kT}V_{CB}\right)\right] \tag{2.66}$$

where G_b is defined by

$$G_b \equiv \int_0^W N_B(x_B)\,dx_B \tag{2.67}$$

and is called the *base Gummel number* [6].

Repeating the foregoing analysis for the hole current in the emitter, we obtain

$$J_{pE}\int_0^{l_E} \frac{dx_E}{p_{Eo}(x_E)} = qD_{pE}\left[\frac{p_E(0)}{p_{Eo}(0)} - \frac{p_E(l_E)}{p_{Eo}(l_E)}\right] \tag{2.68}$$

as the counterpart of (2.64). The minority-carrier concentration $p_{Eo}(x_E)$ is described by an equation similar to (2.65):

$$p_{Eo}(x_E) = \frac{n_{iE}^2}{N_E(x_E)} \tag{2.69}$$

where n_{iE} is the intrinsic carrier concentration in the emitter. Because the bandgap E_g of silicon starts decreasing as the doping concentration exceeds approximately 10^{18} cm^{-3}, most of the emitter has a narrower bandgap E_{gE} in comparison with the E_g of the rest of the BJT regions. According to (1.40), the so-called *bandgap narrowing* $\Delta E_g \equiv E_g - E_{gE}$ causes n_{iE} to be greater than n_i as expressed by

$$n_{iE} = n_i \exp(\Delta E_g/2kT) \tag{2.70}$$

Rewriting (2.68) with (2.69) and (2.70), and using the boundary values

$$p_E(0) = p_{Eo}(0) \exp\left(-\frac{q}{kT}V_{EB}\right)$$

$$p_E(l_E) = p_{Eo}(l_E)$$

we finally obtain

$$J_{pE} = \frac{qD_{pE}n_i^2}{G_e}\left[\exp\left(-\frac{q}{kT}V_{EB}\right) - 1\right] \tag{2.71}$$

where

$$G_e \equiv \int_0^{l_E} N_E(x_E) \exp\left[-\frac{\Delta E_g(x_E)}{kT}\right] dx_E \tag{2.72}$$

is an *effective* emitter Gummel number. Since ΔE_g is positive, G_e is smaller than the total emitter doping concentration per unit area, which is defined by

$$G_e' \equiv \int_0^{l_E} N_E(x_E)\, dx_E \tag{2.73}$$

and is commonly referred to as the *emitter Gummel number.*

Comparing (2.66) with (2.28), and (2.71) with (2.22), clearly shows that $N_B W$ and $N_E l_E$ must be replaced respectively by G_b and G_e in the case of nonuniform base and emitter doping profiles. This conclusion generalizes the Ebers-Moll equations, (2.44) through (2.46) or (2.52) through (2.54), for any given doping concentration profile in these two regions provided that the appropriate Gummel number is substituted for N_BW and N_El_E in the expressions previously obtained for I_S, β_F, and β_R. For example, for the typical bias conditions $V_{BE} \gg kT/q$ and $V_{CB} \gg kT/q$ of the forward active mode, these parameters are described by

$$I_S = \frac{Aqn_i^2 D_{nB}}{G_b} \tag{2.74}$$

$$\beta_F = \frac{D_{nB}G_e}{D_{pE}G_b} \tag{2.75}$$

$$\beta_R = \frac{1}{\dfrac{D_{pC}G_b}{D_{nB}L_{pC}N_C} + \dfrac{W_{tC}G_b}{2n_i\tau_{or}D_{nB}}} \tag{2.76}$$

2.2.7 The Early Effect

Note from (2.74) through (2.76) that all three BJT parameters I_S, β_F, and β_R are inversely proportional to the base Gummel number G_b. Being defined as the net doping concentration per unit area in the base bulk region, the base Gummel number G_b, is a function of the two bias voltages V_{BE} and V_{CB} because the base-side widths of the two transition regions can be modulated by these voltages. We, therefore, expect I_S, β_F, and β_R to be bias dependent through G_b. This dependence is known as the *Early effect* [7]. Since it is most influential in the forward active mode, we will examine its effect specifically for that mode.

First of all, note from (2.61) and (2.62) that the Early effect has no influence on I_B because the ratios I_S/β_F and I_S/β_R are independent of G_b. In studying the effect on I_C, on the other hand, we can ignore the last term on the right-hand side of (2.62) because it is much smaller than the first term for typical values of V_{BE} in the forward active mode. Therefore, we can attribute the effect on I_C to the modulation of I_S only.

For a quantitative modeling of the Early effect, note from Figure 2.9(a) that the base Gummel number can be expressed as

$$G_b = G_{bm} - G_{be} - G_{bc} \tag{2.77}$$

where G_{bm} is the net doping concentration per unit area of the *metallurgical base* extending between the two junction interfaces, and G_{be} and G_{bc} are the similarly defined concentrations on the base sides of the EB and CB transition regions. In the forward active mode, G_{be} is small because the forward bias reduces the barrier height and, hence, the field and the charge sustained by the field in the EB transition region. More importantly, the bias level, and therefore G_{be}, do not change much

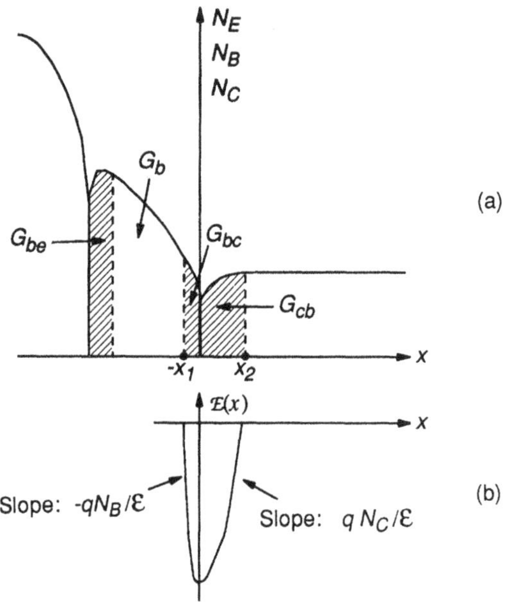

Figure 2.9 The variables involved in the analysis of the Early effect: (a) areal charge densities and (b) electric-field profile in the CB transition region.

in the forward active mode. The situation in the CB transition region, however, is markedly different. The barrier height, and therefore the field, are enhanced by a reverse bias. As a result G_{bc} is larger than its equilibrium value and, more importantly, can vary considerably because V_{CB} has a much wider range of variation in comparison with V_{BE} in the forward active mode. Essentially, therefore, the Early effect in the forward active mode is due to the variation of G_{bc} with V_{CB}.

Selecting the interface of the CB junction as the origin of a position variable x, as shown in Figure 2.9, we can express G_{bc} as

$$G_{bc} \equiv \int_{-x_1}^{0} N_B(x)\, dx \tag{2.78}$$

where $x = -x_1$ is the location of the base-side boundary of the CB transition region. As we discussed in Section 2.1, the base side of the transition region, in which G_{bc} is defined, is negatively charged because it contains a hole concentration of less than $N_B(x)$ and an electron concentration of more than $n_i^2/N_B(x)$. Assuming both concentrations to be below $N_B(x)$, the charge density in the region $-x_1 < x < 0$ can be expressed as

$$\rho(x) \cong -qN_B(x), \qquad \text{for } -x_1 < x < 0 \tag{2.79}$$

for which Poisson's equation (1.3) indicates a negative field profile of a gradient $d\mathscr{E}/dx = -qN_B/\epsilon$. Also note the assumption of a negligible field in the base bulk. Moving to the collector side of the transition region, we can define the doping concentration per unit area as

$$G_{cb} \equiv \int_0^{x_2} N_C\, dx \cong N_C x_2 \tag{2.80}$$

where $x = x_2$ is the location of the transition region boundary on the collector side. Again assuming negligibly small carrier concentrations, the charge density on this side can be expressed as

$$\rho(x) \cong qN_C, \qquad \text{for } 0 < x < x_2 \tag{2.81}$$

which, since it is positive, leads to a field profile of a positive gradient $d\mathscr{E}/dx = qN_C/\epsilon$ as shown in Figure 2.9(b). Since the field vanishes at $x = -x_1$ and at $x = x_2$, the space charge between these two boundaries must retain its dipole property, that is,

$$\int_{-x_1}^{0} \rho\, dx = -\int_0^{x_2} \rho\, dx$$

Substituting (2.79) and (2.81) into this equation, and comparing with (2.78) and (2.80), we reach the conclusion

$$G_{bc} = G_{cb} = N_C x_2 \tag{2.82}$$

Quite obviously, the variation of G_{bc} with V_{CB} must be implicit in x_2 because N_C is independent of V_{CB}. To determine how x_2 depends on V_{CB}, let us apply (1.6) to the CB transition region:

$$\psi(x_2) - \psi(-x_1) = -\int_{-x_1}^{x_2} \mathscr{E}(x)\, dx$$

But $\psi(x_2) - \psi(-x_1)$, being the potential barrier across the region, equals $V_{CB} + V_{b(CB)}$. In other words, the area bounded by the field profile of Figure 2.9(b) can be equated to $-[V_{CB} + V_{b(CB)}]$. The area bounded on the base side is negligibly small in comparison to the approximately triangular area bounded on the collector side because N_B is much greater than N_C. Therefore, $-[V_{CB} + V_{b(CB)}] \cong \mathscr{E}(0)x_2/2$. Combining this with $\mathscr{E}(0)/x_2 = -qN_C/\epsilon$, as implied by the field gradient, we obtain

$$x_2 = \sqrt{\frac{2\epsilon}{qN_C}(V_{CB} + V_{b(CB)})} \tag{2.83}$$

Substituting this equation for x_2 in (2.82) yields the following relationship between G_{bc} and V_{CB}:

$$G_{bc} = \sqrt{\frac{2\epsilon N_C}{q}(V_{CB} + V_{b(CB)})} \tag{2.84}$$

It is obvious from (2.84), (2.77), (2.74), and (2.62) that an increasing collector-base bias V_{CB} makes G_{bc} larger, G_b smaller, I_S larger, and, consequently, I_C larger. This effect on I_C is usually represented by the *Early voltage* parameter V_{AF}. As shown in Figure 2.10, V_{AF} can be determined by extrapolating to the V_{CB} axis the slope of the $I_C - V_{CB}$ characteristic at a given operating point of fixed V_{CB}, for example, V_{CBQ} and I_{CBQ}.

To relate V_{AF} to structural parameters, let us rewrite (2.62) for a fixed V_{BE} using (2.74) and (2.77), and assuming $G_{be} \ll G_{bm}$:

$$I_C = \frac{\text{constant}}{G_{bm} - G_{bc}}$$

whose derivative with respect to V_{CB} is

$$\left.\frac{dI_C}{dV_{CB}}\right|_{V_{CB}=V_{CBQ}} = \frac{I_{CQ}}{G_{bm} - G_{bc}} \left.\frac{dG_{bc}}{dV_{CB}}\right|_{V_{CB}=V_{CBQ}} \tag{2.85}$$

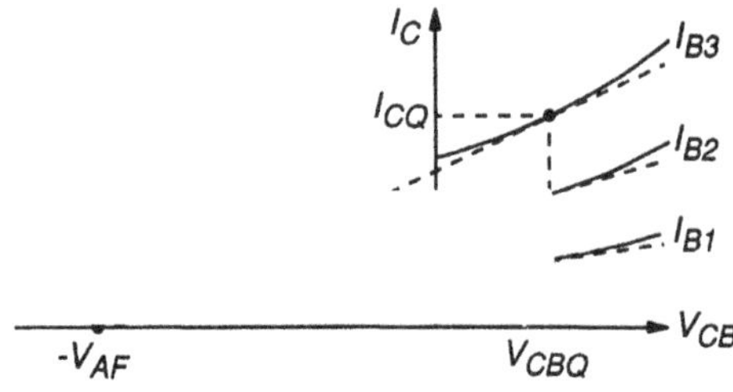

Figure 2.10 Representing the Early effect with an Early voltage V_{AF}.

Substituting (2.84) for G_{bc}, equating the derivative to $I_{CQ}/(V_{CBQ} + V_{AF})$ in accordance with the definition of V_{AF} (see Figure 2.10), and assuming a V_{CBQ} of at least a few volts so that $V_{b(CB)}$ is negligible in comparison with V_{CBQ}, we obtain

$$V_{AF} = G_{bm}\left(\frac{2qV_{CBQ}}{\epsilon N_C}\right)^{1/2} - 3V_{CBQ} \tag{2.86}$$

As will be shown in Section 2.4.2, the small-signal output resistance of the BJT is proportional to V_{AF}, which is why a large V_{AF} is a desirable property. It is obvious from (2.86) that to increase V_{AF}, and thus to make the Early effect less effective, one must increase the base Gummel number and/or decrease the collector doping concentration.

2.3 HIGH-BIAS EFFECTS

The analyses and models presented in previous sections are based on a number of assumptions including one-dimensionality, low-level injection in all three bulk regions and an absence of any generation process other than the thermally activated one. Some of these assumptions limit the validity of the primary BJT model to a relatively small range of bias voltages and terminal currents. Since the bias range involved in practice may exceed the limitations of the primary model, it is necessary to revise the latter in order to be able to predict device behavior in response to such large-bias excitations. In this section, we will analyze the most significant high-bias effects, and will predict their impact on the terminal characteristics of the BJT.

2.3.1 Base Resistance and Emitter Current Crowding

In Section 2.2.1, we based the foundations of the basic BJT model on an assumption of negligible bulk voltages. As can be understood from (2.11), this assumption becomes less justifiable at high levels of current density. For the emitter bulk, whose length is very small and doping concentration very high, it usually takes an impractically high emitter current density to develop an appreciable bulk voltage. Therefore, the basic assumption of a negligible emitter bulk voltage is justifiable for all practical bias levels. In contrast to the emitter in this respect is the collector bulk, whose length is the largest and doping concentration the lowest among all bulk regions. A significant collector bulk voltage can easily develop at relatively low collector current density levels and can lead to a considerable deviation from the basic BJT operation, as is discussed in Section 2.3.2. The base region is doped to an intermediate concentration and supports a negligible majority-carrier current, hence voltage, along the direction of main current flow between emitter and collector. However, the actual base current flows along a transversal path of significant length and of very narrow cross-sectional area. As we now discuss in detail, this can result in a considerable voltage drop in the transversal direction, which, in turn, can significantly alter device behavior at high bias levels [8–10].

Physical Origins

We know from Section 2.2.5 that I_{pE} is the dominant component of the base current in the forward active mode. As shown in Figure 2.11, this current is carried by hole transport from base contact into the active base region where injection to the emitter takes place.

Along the *inactive base region,* which is located between points B and B', the cross-sectional area through which I_B flows is relatively large, and therefore the current density is small. Combined with the relatively high base doping concentration, this should result in a relatively small voltage drop between these two points, as can be deduced from (2.11). In other words, the *resistance* of the inactive base is relatively small. However, once the holes enter the active base region, they are confined to a much narrower cross-sectional area, which implies a much higher current density, hence a larger voltage drop in the transversal direction between points B' and B''. On the emitter side of the junction, however, the potential is uniformly distributed along the same direction because (1) there is no significant transversal current flow in the emitter and (2) the emitter has negligible resistance as a result of a very high doping concentration. Therefore, the internal bias across the junction at point B' becomes larger than the bias at point B''. Now, recall that the collector current is a result mainly of an electron injection from the emitter into the base, and that the resulting current density J_{nB} is an exponentially increasing function of the base-emitter voltage. The above-mentioned nonuniformity in the internal bias will, therefore, cause J_{nB} to be larger at point B' than at point B''! Of course, the larger the I_B, the bigger the transversal voltage drop, and thus the more pronounced the *crowding* of J_{nB} at the contact side of the active base.

To understand the effect of the transversal voltage and current crowding on the collector current, we need to remember that the latter is the surface integral

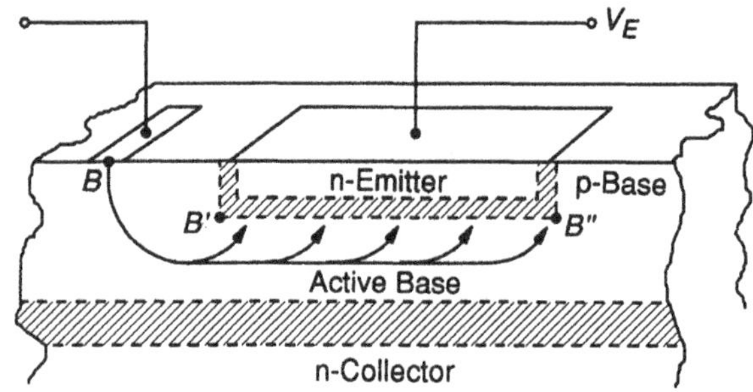

Figure 2.11 Hole current conduction in the base region of a BJT.

of J_{nB} over the active base area. For a given external base-emitter voltage, therefore, J_{nB} will be less than what is predicted by the primary BJT model because the actual internal base-emitter bias is lesser than the externally applied base-emitter voltage. This implies an $I_C - V_{BE}$ Gummel slope that is smaller than what is predicted by the primary model.

A Mathematical Model for Base Resistance

We will now derive a mathematical model for the resistance of the active base region assuming a single base contact as shown in Figure 2.12, where V and I denote the voltage and transversal current in the active base region, respectively. Since I is carried by the holes drifting along y, we can write[1]:

$$I = -q\mu_{pB}LG_b\frac{dV}{dy} \tag{2.87}$$

Conversely, the infinitesimal variation of I along dy is given by $dI = -LJ_{pE}dy$. Using (2.71) for J_{pE} and replacing V_{EB} with $V_E - V$ yield

$$\frac{dI}{dy} = -\frac{qD_{pE}n_i^2L}{G_e}\exp\left(-\frac{q}{kT}V_E\right)\exp\left(\frac{q}{kT}V\right) \tag{2.88}$$

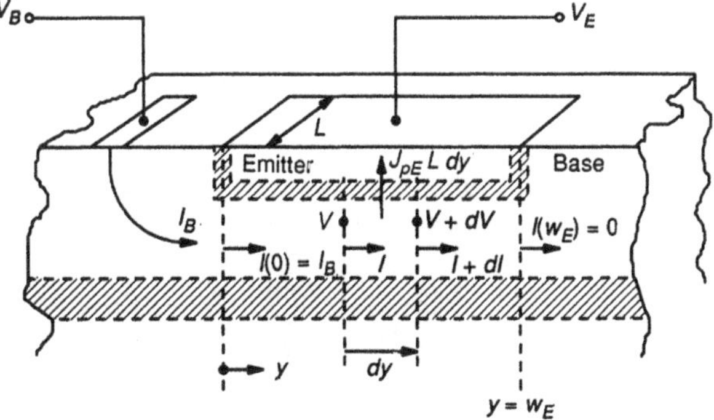

Figure 2.12 The variables involved in the analysis of the base resistance and the emitter current crowding effect.

---

[1]Take a hole drift current density expression $J = q\mu_{pB}p_B\mathscr{E}$; substitute $-dV/dy$ for $\mathscr{E}$; multiply by Ldx to find dJ; integrate over x between 0 and W assuming an x-independent dV/dy; and replace $\int_0^W p_B dx$ with G_b because $p_B = N_B$.

From (2.87) and (2.88), we obtain

$$\frac{d^2I}{dy^2} + \frac{I}{qD_{pB}LG_b}\frac{dI}{dy} = 0 \tag{2.89}$$

whose solution with $I(0) = I_B$ (because the entire base current enters the active base at $y = 0$) and $I(W_E) = 0$ (because no current flows beyond $y = W_E$) is given by

$$I(y) = 2\frac{qD_{pB}LG_b}{W_E}Z \tan\left[Z\left(1 - \frac{y}{W_E}\right)\right] \tag{2.90}$$

where

$$Z \tan Z = \frac{W_E}{2qD_{pB}LG_b}I_B \tag{2.91}$$

Finding the spatial derivative of $I(y)$ from (2.90), using it in (2.88) for $y = 0$, and replacing V with V_B, we obtain the following equation[2]:

$$V_{BE} = \frac{kT}{q} \ln\frac{2D_{pB}G_bG_eZ^2}{D_{pE}n_i^2W_E^2 \cos^2 Z} \tag{2.92}$$

which relates V_{BE} to I_B through Z in terms of various structural parameters. It is important to note that V_{BE} is the externally applied bias. Therefore, (2.92) defines the EB port characteristic and replaces the equation $V_{BE} = (kT/q)\ln(\beta_F I_B/I_S)$ of the primary model [see(2.61)].

In practice, it is customary to model the effect of base resistance by connecting a lumped resistance R_B in series with the base of an ideal BJT, as shown in Figure 2.13. The base voltage V'_{BE} of this ideal BJT is then given by

$$V'_{BE} = \frac{kT}{q}\ln\left(\frac{\beta_F}{I_S}I_B\right) = \frac{kT}{q}\ln\frac{G_eI_B}{qn_i^2LW_ED_{pE}} \tag{2.93}$$

Also obvious from Figure 2.13 is the equation $R_B = (V_{BE} - V'_{BE})/I_B$. By substituting (2.92) and (2.93) into this equation, and using (2.91), we obtain

$$R_B = \frac{kT}{q}\frac{1}{I_B}\ln\frac{Z}{\sin Z \cos Z} \tag{2.94}$$

[2]If the resistance between the base contact and $y = 0$ is sufficiently small, one can assume $V(0) = V_B$.

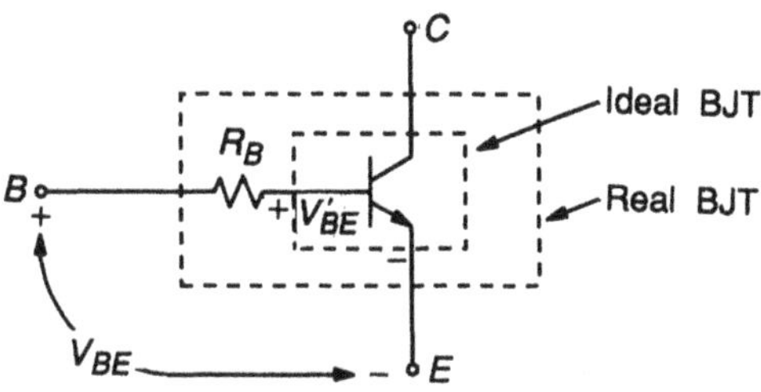

Figure 2.13 Representation of the base resistance in a circuit model.

Shown in Figure 2.14 as a function of I_B is the base resistance as calculated from this equation for a typical BJT. For low levels of base current, the resistance is virtually constant. This so-called *zero-bias base spreading resistance* is given by

$$R_B = \frac{1}{3}\frac{kT}{q}\frac{W_E}{qD_{pB}LG_b} \tag{2.95}$$

For large levels of base current, for which Z approaches $\pi/2$, (2.94) reduces to

$$R_B = \frac{kT}{q}\frac{1}{I_B}\ln\frac{W_E I_B}{2qD_{pB}LG_b} \tag{2.96}$$

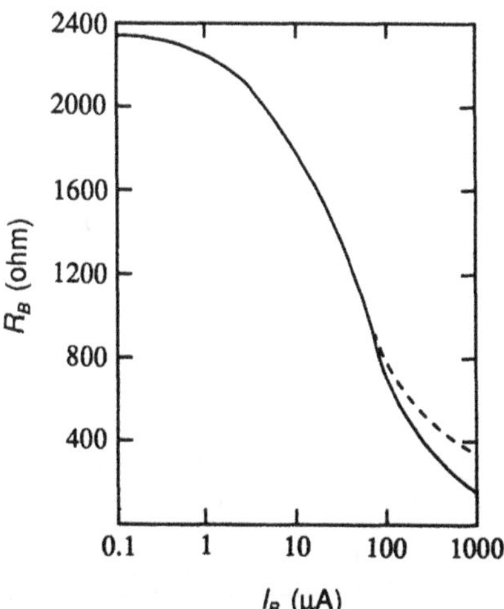

Figure 2.14 The variation of base resistance with base current as calculated from (2.94) for a typical BJT.

and thus becomes a decreasing function of I_B. This bias dependence is a consequence of the current crowding effect, which confines the transversal transportation of holes to a fraction of the emitter width W_E. It is important to note that (2.94) through (2.96) describe the active base resistance, which is more popularly known as the *intrinsic base resistance*. At sufficiently high current levels, this resistance decreases below the *extrinsic base resistance,* which is the bias-independent resistance between the contact and the active base. The total base resistance will, therefore, approach the extrinsic base resistance at very high bias levels, and thus once more become bias independent as shown by the dashed tail in Figure 2.14.

The base resistance has little, if any, effect on the dc terminal characteristics at low bias levels because the voltage drop across this resistance is negligible. At high bias levels, however, it has a significant effect. To see how, note from (2.91) that, for a Z approaching $\pi/2$, we can write $Z \tan Z \cong Z/\cos Z$. Squaring the right-hand side of (2.91), substituting it for $Z^2/\cos^2 Z$ in (2.92), and rearranging, we obtain the following expression for I_B as a function of V_{BE} for high bias levels:

$$I_B = qLn_i\sqrt{2D_{pB}D_{pE}\frac{G_b}{G_e}}\exp\left(\frac{q}{2kT}V_{BE}\right) \qquad (2.97)$$

Obviously, the rate at which I_B varies with V_{BE} on the semilog Gummel plot is halved due to the current-dependent base resistance. Since β_F is unaffected, the collector current Gummel plot exhibits a similar reduction of slope.

Equations (2.90) through (2.97) have been derived for the single base contact structure depicted in Figure 2.11. In many cases, the base is contacted on both sides of the emitter. In such cases (2.89) must be solved with the boundary conditions $I(0) = I_B/2$ and $I(W_E/2) = 0$. This results in

$$R_B = \frac{1}{12}\frac{kT}{q}\frac{W_E}{qD_{pB}LG_b} \qquad (2.98)$$

for low bias levels, and in

$$R_B = \frac{kT}{q}\frac{1}{I_B}\ln\frac{W_E I_B}{8qD_{pB}LG_b} \qquad (2.99)$$

for high bias. A comparison between these equations and (2.95-96) clearly shows how R_B can be reduced by adding a second base contact. Also obvious from these equations is the significance of selecting a small emitter aspect ratio W_E/L. This is why the emitter is usually patterned into stripes in medium and high power BJTs.

2.3.2 High-Level Injection Effects

Foundations

According to (1.64), low-level injection conditions prevail only if excess concentrations are negligible in comparison with the doping concentration. This condition is definitely met in the cutoff mode of operation because the magnitude of the excess concentrations cannot exceed the minority-carrier equilibrium concentration, which itself is negligible in comparison with the doping concentration. On the other hand, excess concentrations in the emitter and base regions in the forward active mode, and in all three regions in saturation, are exponentially increasing functions of the forward-biasing voltages as indicated by (2.19) and (2.21) for the emitter, by (2.25), (2.26), and (2.27) for the base, and by (2.30) and (2.31) for the collector. Therefore, the assumption of low-level injection should cease to be justifiable for sufficiently high levels of bias. A remodeling of the BJT is therefore necessary to explain device behavior at such high bias levels. Considering the fact that high precision in device modeling is meaningful mostly in analog circuit applications, in which the device is supposed to operate only in the forward active mode, one is tempted to remodel only the bulk regions bordering the EB junction because one expects V_{CB} to reverse bias the CB junction, thus keeping the collector side of the base and the collector itself under low-level injection conditions. This would be a false expectation because, at sufficiently high levels of J_C, the voltage drop along the active collector region can become larger than the externally applied V_{CB} and, therefore, can forward bias the CB junction. As a result, the neighboring bulk regions of this junction can be forced into high-level injection conditions as well. For this reason, we cannot be presumptuous about the conditions prevailing in the base and collector regions, but because the excess concentrations are inversely proportional to the doping concentration, we can safely assume low-level injection conditions in the two very heavily doped bulk regions of the emitter and buried layer. Clearly, the base and collector bulks enter high-level injection at much lower bias levels than these two regions, and thus dominate the high-bias behavior of the device. On the basis of these considerations, we will now remodel the BJT.

As understood from the structural schematic given in Figure 2.15(a), the two dimensionality of the base region will be taken into account in developing the model. The corresponding energy-band diagram, given in Figure 2.15(b), contains two branches to the left of $x_B = W$, one belonging to the main current-flow axis between points E and B', and the other belonging to the path B-B'' along the inactive base region. It is assumed that points B' and B'' have similar band prop-

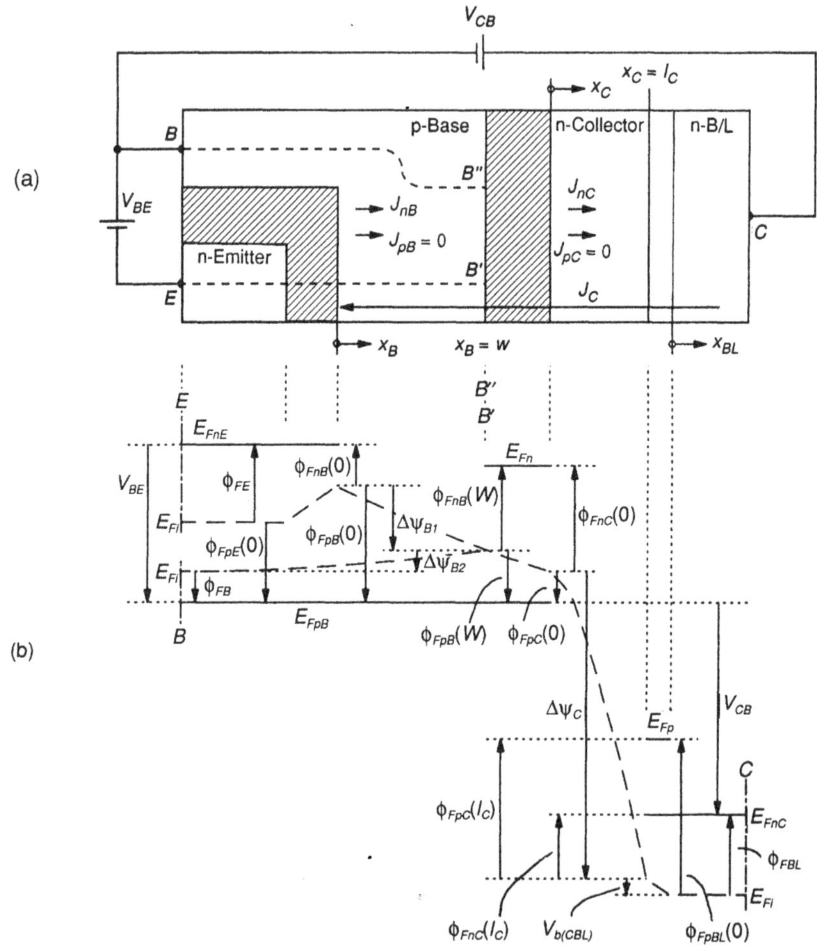

Figure 2.15 The variables involved in the analysis of high-level injection effects in the base and the collector: (a) structural schematic and internal variables and (b) energy-band diagram.

erties, which is justifiable if the current crowding effect is negligible. Two other fundamental assumptions of the model are discussed next.

1. Apart from the EB and CB transition regions, everywhere in the structure a condition of quasineutrality prevails, that is, $p - n + N \cong 0$. This assumption was proved in Section 1.3.3 to be valid under low-level injection conditions and

is now being extended to the case of high-level injection as well. By arranging Poisson's equation as

$$\frac{\epsilon}{q}\frac{d\mathscr{E}}{dx} - p + n - N = 0$$

we recognize that, as long as $(\epsilon/q)(d\mathscr{E}/dx)$ is sufficiently smaller than p or n or N, the condition of quasineutrality is satisfied regardless of the level of injection. One can apply this test once the field and carrier concentration profiles have been obtained on the basis of quasineutrality; therefore, the level of bias at which this assumption is no longer justifiable can be determined.

2. The three terminals at the surfaces of the emitter, base, and collector regions are assumed to create ohmic contacts, where excess carrier concentrations vanish. This is why the electron and hole quasi-Fermi levels merge into a single Fermi level at points E, B, and C. The potential difference between this unified Fermi level and the intrinsic Fermi level is equal to the equilibrium Fermi potential, that is, ϕ_{FE} at the emitter contact, ϕ_{FB} at the base contact and ϕ_{FBL} at the buried-layer contact. The external bias voltages V_{BE} and V_{CB} define the separation between these three Fermi levels at the contacts. Also notice the position-independent majority-carrier quasi-Fermi levels in the emitter and buried-layer bulk regions. This position independence is a consequence of the assumption of low-level injection in these regions. The majority-carrier quasi-Fermi level in the base is also position-independent as can be justified from (1.107) considering the large hole concentration and weak hole current density associated with this region.

Remodeling the Base

Under high-level injection conditions, the field in the base may become large enough to create a potential difference $\Delta\psi_{B1}$ along the main axis of current flow. If point B'' is also under high-level injection, we expect a potential difference $\Delta\psi_2$ to develop between points B'' and B as well, even if the base current, which more or less follows this path, is still negligible. This is why both branches of the energy-band diagram of the base section are bent in Figure 2.15(b).

We now rewrite the two current density equations in their exact form for $0 \leq x_B \leq W$:

$$J_{nB} = q\mu_{nB}n_B\mathscr{E}_B + qD_{nB}\frac{dn_B}{dx_B} \tag{2.100}$$

$$J_{pB} = q\mu_{pB}p_B\mathscr{E}_B - qD_{pB}\frac{dp_B}{dx_B} \tag{2.101}$$

As we will show when remodeling the collector, the hole current in the collector, and therefore at $x_B = W$, is negligibly small. If a small base current, hence a small hole current at $x_B = 0$, is also assumed, we can use $J_{pB} \cong 0$ in (2.101). Besides, the assumption of quasineutrality allows us to substitute $n_B + N_B$ for p_B in the same equation whose solution for $\mathscr{E}_B$ will therefore yield

$$\mathscr{E}_B = \frac{kT}{q} \frac{1}{n_B + N_B} \frac{dn_B}{dx_B} \tag{2.102}$$

Substituting this into (2.100) for $\mathscr{E}_B$, replacing J_{nB} with $-J_C$, and rearranging, we obtain the differential equation

$$\frac{J_C}{qD_{nB}} dx_B = -\frac{2n_B + N_B}{n_B + N_B} dn_B$$

whose integration between $x_B = 0$ and $x_B = W$ for a uniform N_B yields

$$J_C = \frac{qD_{nB}}{W}\left\{2[n_B(0) - n_B(W)] - N_B \ln\left[1 + \frac{n_B(0) - n_B(W)}{N_B + n_B(W)}\right]\right\} \tag{2.103}$$

One can easily show with a series approximation of the logarithmic term that,

B B

(2.103) can be simplified to

$$J_C = \frac{qD_{nB}}{W}[n_B(0) - n_B(W)]$$

which indeed is what the basic low-level-injection-based model predicts for the collector current. At very high levels of injection, on the other hand, (2.103) predicts

$$J_C = 2\frac{qD_{nB}}{W}[n_B(0) - n_B(W)] \tag{2.104}$$

Note from this equation that, for a given electron concentration gradient $[n_B(0) - n_B(W)]/W$, the current doubles for high-level injection in base. To understand the physics of this phenomenon, refer to (2.100). Under low-level injection conditions, the minority-carrier electrons move predominantly by diffusion in accordance with the second term on the right-hand side of (2.100). As the level of injection increases, an electric field develops in the region in accordance with (2.102). This field, because it is directed toward the emitter, aids the electron movement by enhancing the drift component, which is represented by the first term on the right-hand side of (2.100). At extremely high levels of injection, at which $n_B \gg N_B$, this drift component equals the diffusion component, as one can determine from (2.102) and (2.100). This is the ultimate situation represented by (2.104).

Let us now determine $n_B(0)$, which is quite generally described by (1.105) as follows:

$$n_B(0) = n_i \exp\left[-\frac{q}{kT}\phi_{FnB}(0)\right] \tag{2.105}$$

For $\phi_{FnB}(0)$, we can write the following loop equation from the energy-band diagram:

$$\phi_{FnB}(0) = -V_{BE} + \phi_{FB} + \Delta\psi_{B2} + \Delta\psi_{B1} \tag{2.106}$$

where $\Delta\psi_{B1}$ can be determined by integrating (2.102) in accordance with (1.6). The result is

$$\Delta\psi_{B1} = \frac{kT}{q}\ln\frac{N_B + n_B(0)}{N_B + n_B(W)}$$

To formulate $\Delta\psi_{B2}$, we can write an expression similar to (2.102) for the field along the path B''-B provided the base current is small. Integrating the expression along the same path, we obtain

$$\Delta\psi_{B2} = \frac{kT}{q}\ln\frac{N_B + n_B(W)}{N_B + n_{Bo}} \cong \frac{kT}{q}\ln\frac{N_B + n_B(W)}{N_B}$$

Substituting these two equations together with

$$\phi_{FB} = \frac{kT}{q}\ln\frac{N_B}{n_i}$$

into (2.106), using the latter in (2.105), and solving for $n_B(0)$, we find

$$n_B(0) = -\frac{N_B}{2} + \sqrt{\left(\frac{N_B}{2}\right)^2 + n_i^2\exp\left(\frac{q}{kT}V_{BE}\right)} \tag{2.107}$$

Under low-level injection conditions, for which $n_B(0)$ is much smaller than N_B, (2.107) indeed reduces to (2.26) of the basic BJT model. For an extremely high level of injection, however, it approaches the form

$$n_B(0) = n_i \exp\left(\frac{q}{2kT}V_{BE}\right)$$

whose comparison with (2.26) indicates a high-injection effect that halves the rate of change of log $n_B(0)$ with V_{BE}. This reduction in slope is due to the fact that the voltage imposed on the EB transition region, which determines $n_B(0)$, falls increasingly behind the external bias V_{BE} because the base bulk voltage $\Delta\psi_{B1} + \Delta\psi_{B2}$ grows with the increasing level of injection.

The increasing rate of variation of J_C with the electron gradient and the decreasing rate of variation of log $n_B(0)$ with V_{BE} are the two main high-injection effects associated with the base region. These two effects, in combination, are generally known as the *Webster effect* [11].

In the conventional forward active mode, in which $n_B(W) \cong 0$, the Webster effect reduces the slope of the J_C-V_{BE} Gummel plot at large values of V_{BE} because, while J_C remains proportional to $n_B(0)$, the rate by which the latter varies with V_{BE} decreases. However, the Webster effect is rarely observed in isolation from other high-bias effects. Later in this section, we will illustrate with a numerical example the way in which the Gummel plot is influenced by a combination of the Webster and quasisaturation effects.

Earlier in this section we stated that, due to the relatively heavy doping concentration involved, the emitter bulk region was still under low-level injection conditions when what we now call the Webster effect prevailed in the base. We thus extended the validity of (2.18) and (2.19) to describe J_{pE} and therefore I_B even for those bias levels for which the base was under the influence of the Webster effect, that is,

$$J_{pE} = q\frac{D_{pE}}{l_E}[p_E(0) - p_{Eo}] \tag{2.108}$$

Of course, the Webster effect is accompanied by an enhancement of the base bulk voltage, which forces the voltage across the EB transition region to fall increasingly behind V_{BE} as the level of injection increases in the base. As a result, (2.21) might appear to be an invalid description of $p_E(0)$ when the Webster effect prevails in the base because this equation was derived under the assumption of a negligible bulk voltage not only in the emitter but also in the base. This raises the question as to whether the low-level injection model of J_{pE}, and therefore of I_B, indeed remains the same under the influence of the Webster effect. To answer this question let us write

$$p_E(0) = n_i \exp\left[\frac{q}{kT}\phi_{FpE}(0)\right]$$

and then substitute for $\phi_{FpE}(0)$ the following loop equation written from the energy-band diagram of Figure 2.15(b),

$$\phi_{FpE}(0) = \phi_{FpB}(0) - \phi_{FnB}(0) + \phi_{FE}$$

where

$$\phi_{FpB}(0) = \frac{kT}{q}\ln\frac{p_B(0)}{n_i} = \frac{kT}{q}\ln\frac{N_B + n_B(0)}{n_i}$$

$$\phi_{FnB}(0) = -\frac{kT}{q}\ln\frac{n_B(0)}{n_i}$$

$$\phi_{FE} = -\frac{kT}{q}\ln\frac{N_E}{n_i}$$

We obtain from these equations the following general expression for $p_E(0)$,

$$p_E(0) = \frac{n_B(0)[N_B + n_B(0)]}{N_E}$$

which turns (2.108) into

$$J_{pE} = q\frac{D_{pE}}{l_E N_E}[N_B + n_B(0)]n_B(0)$$

for $p_{Eo} \ll p_E(0)$. Using (2.107) for $n_B(0)$ in this equation, we finally obtain

$$J_{pE} = \frac{qD_{pE}n_i^2}{N_E l_E}\exp\left(\frac{q}{kT}V_{BE}\right)$$

which is precisely the same as (2.22) for the forward active mode of operation. We therefore conclude that J_{pE}, and therefore I_B, are not influenced by the Webster effect.

Remodeling the Collector

As we mentioned in Section 2.3.1, the collector bulk region can easily absorb a large voltage even at moderate levels of current density as a result of its relatively large length and light doping concentration. For a given positive V_{CB}, an increasing J_C therefore can easily reduce the reverse bias across the CB transition region and, ultimately, can force the junction into a forward bias condition. In consequence, the BJT can exit the forward active mode and enter saturation despite the positive-valued external V_{CB}. This phenomenon, called *quasisaturation*, cannot be treated as a simple saturation mode of operation as defined in the context of the basic model because a high-level injection condition in the collector bulk is usually associated with it [12–15]. Therefore, we need to remodel the collector.

First of all, we will show that the hole current density in the collector is still negligibly small even if high-level injection conditions prevail. For this purpose, let us adopt (2.29) and (2.30) to describe the hole current at the high-low junction boundary $x_{BL} = 0$ of the buried layer, which is assumed to be operating under low-level injection conditions. We can write

$$J_{pBL}(0) = \frac{qD_{pBL}}{L_{pBL}}[p_{BL}(0) - p_{BLo}] \tag{2.109}$$

where

$$p_{BL}(0) = n_i \exp\left[\frac{q}{kT}\phi_{FpBL}(0)\right]$$

A loop equation written from the energy-band diagram with $\phi_{FpBL}(0)$, $V_{b(CBL)}$, and $\phi_{FpC}(l_C)$ yields

$$\phi_{FpBL}(0) = \phi_{FpC}(l_C) - V_{b(CBL)}$$

where $\phi_{FpC}(l_C)$ is described by

$$\phi_{FpC}(l_C) = \frac{kT}{q}\ln\frac{p_C(l_C)}{n_i} = \frac{kT}{q}\ln\frac{n_C(l_C) - N_C}{n_i}$$

The barrier potential $V_{b(CBL)}$ of the high-low junction is assumed to retain its equilibrium value:

$$V_{b(CBL)} \cong \frac{kT}{q}\ln\frac{N_{BL}}{N_C} \tag{2.110}$$

From the last four equations, we obtain

$$p_{BL}(0) = \frac{N_C}{N_{BL}}[n_C(l_C) - N_C]$$

On the other hand, (2.110) also implies $n_C(l_C) \cong N_C$, hence, $p_{BL}(0) \cong 0$. Using this result in (2.109), we conclude that J_{pBL} at $x_{BL} = 0$ is indeed small. Ignoring generation and recombination between $x_C = 0$ and $x_{BL} = 0$, we can assume a position-independent J_{pC}, which, therefore, must equal the negligibly small $J_{pBL}(0)$. Having reached this conclusion, we can now assume $J_C = -J_{nC}$, where J_{nC} is the electron current density in the collector. Adopting the general electron current density equation (1.18) for the latter, we can write

$$J_C = -q\mu_{nC}\left(n_C\mathscr{E}_C + \frac{kT}{q}\frac{dn_C}{dx_C}\right) \tag{2.111}$$

for the collector current density. A few words about the electron mobility appearing in this equation are in order. Considering that the potential difference $\Delta\psi_C$ across the active collector region can be as large as the external bias V_{CB} (and be even larger under quasisaturation conditions) and that V_{CB} could be several tens of volts in practice, the model must allow for a large $\Delta\psi_C$ and therefore a large field $\mathscr{E}_C$ in the active collector. For this reason, it is appropriate to take into consideration the degrading effect of the electric field on mobility. This effect, which is briefly mentioned at the end of Section 1.1.3 and is illustrated in Figure 1.7, can be represented by the following first-order empirical expression:

$$\mu_{nC} \cong \frac{v_s}{\mathscr{E}_o - \mathscr{E}_C} \tag{2.112}$$

where v_s is the magnitude of electron saturation velocity, and $\mathscr{E}_o$ is a positive constant called the *critical field*. Also note that $-\mathscr{E}_C$ represents the magnitude of the negatively valued electric field in the active collector region.

An expression for the field $\mathscr{E}_C$ appearing in (2.111) and (2.112) can be obtained by substituting $J_p = 0$ in the general hole current density equation (1.13):

$$\mathscr{E}_C = \frac{kT}{q}\frac{1}{p_C}\frac{dp_C}{dx_C} \tag{2.113}$$

Now, using this for $\mathscr{E}_C$ in (2.111) and (2.112) and substituting (2.112) into (2.111) for μ_{nC} and $n_C = p_C + N_C$ of the quasineutrality condition for n_C, we arrive at the differential equation

$$dx_C = \frac{kT}{q}\frac{1}{J_C\mathscr{E}_o}\frac{J_C - qv_s(2p_C + N_C)}{p_C}dp_C \tag{2.114}$$

Integrating this equation over the collector region and rearranging, we obtain

$$\frac{J_C}{2qv_s}\left[\mathscr{E}_o l_C + \frac{kT}{q}\ln\frac{p_C(0)}{p_C(l_C)}\right] = \frac{kT}{q}[p_C(0) - P_C(l_C)] + \frac{N_C}{2}\frac{kT}{q}\ln\frac{p_C(0)}{p_C(l_C)}$$

Another independent equation is obtained by integrating (2.113) over the same region:

$$\Delta\psi_C = \frac{kT}{q}\ln\frac{p_C(0)}{p_C(l_C)} \tag{2.115}$$

whose substitution into the previous equation yields

$$J_C = \frac{2qv_s}{\mathscr{E}_o l_C + \Delta\psi_C}\left\{\frac{kT}{q}[p_C(0) - p_C(l_C)] + \frac{N_C}{2}\Delta\psi_C\right\} \tag{2.116}$$

A third equation is obtained from the loop involving V_{CB}, ϕ_{FBL}, $V_{b(CBL)}$, $\Delta\psi_C$, and $\phi_{FpC}(0)$ in the energy-band diagram:

$$\Delta\psi_C = V_{CB} + \phi_{FpC}(0) - \phi_{FBL} - V_{b(CBL)}$$

Substituting into this equation

$$\phi_{FBL} = -\frac{kT}{q}\ln\frac{N_{BL}}{n_i}$$

for ϕ_{FBL}, (2.110) for $V_{b(CBL)}$, and

$$\phi_{FpC}(0) = \frac{kT}{q}\ln\frac{p_C(0)}{n_i}$$

for $\phi_{FpC}(0)$, we transform the equation into

$$\Delta\psi_C = V_{CB} + \frac{kT}{q}\ln\frac{p_C(0)N_C}{n_i^2} = V_{CB} + \frac{kT}{q}\ln\frac{p_C(0)}{p_{Co}} \tag{2.117}$$

Equating the right-hand side of (2.115) and (2.117) we arrive at

$$p_C(l_C) = p_{Co}\exp\left(-\frac{q}{kT}V_{CB}\right)$$

which indicates a negligible hole concentration at $x_C = l_C$ for any practical value of V_{CB}. Therefore, (2.116) can be simplified into

$$J_C = \frac{2qv_s}{\mathscr{E}_o l_C + \Delta\psi_C}\left[\frac{kT}{q}p_C(0) + \frac{N_C}{2}\Delta\psi_C\right] \tag{2.118}$$

which, together with (2.117), models current conduction in the active collector region. Shown in Figure 2.16 is the variation of $\Delta\psi_C$ and $p_C(0)$ with J_C as calculated from these two equations for a BJT of $N_C = 8 \times 10^{15}\ \text{cm}^{-3}$ and $l_C = 10\ \mu\text{m}$ operating with $V_{CB} = 4\text{V}$. At relatively low levels of J_C, $p_C(0)$ is many orders of magnitude smaller than p_{Co} as already predicted by the basic model. The value of $\Delta\psi_C$ is also smaller than the external bias V_{CB}; therefore, the voltage imposed on the CB transition region is positive and the junction is reverse biased. However, $\Delta\psi_C$ increases with J_C in accordance with

$$\Delta\psi_C \cong \frac{\mathscr{E}_o l_C J_C}{q v_s N_C - J_C}$$

as one can derive from (2.118) for $(kT/q)p_c(0) \ll (\Delta\psi_C/2)N_C$. This is an ohmic voltage resulting from the nonzero resistance of the active collector region. The resistance involved can be expressed from the last equation as

$$R_C = \frac{\mathscr{E}_o l_C}{q v_s N_C - J_C} \frac{1}{A} \tag{2.119}$$

where A is the collector area. Note that R_C is an increasing function of J_C as a result of the mobility degradation effect. This is why the variation of $\Delta\psi_C$ with J_C is supralinear as depicted in Figure 2.16. As indicated by (2.117), $p_C(0)$ is exponentially dependent on $\Delta\psi_C$ and therefore rises with J_C faster than the exponential rate of (2.117).

At about $J_C = 4.5\ \text{kA/cm}^2$, $\Delta\psi_C$ reaches V_{CB}, and thus removes the reverse bias across the CB transition region. At the same level of bias, $p_C(0)$ reaches p_{Co} as one would expect from an unbiased junction. A further increase in J_C forces the junction into forward bias conditions. The trends established under low bias for $p_C(0)$ and $\Delta\psi_C$, however, continue until $p_C(0)$ reaches the doping level N_C. Thereafter high-level injection conditions are imposed on the collector region. This can be regarded as the onset of the quasisaturation effect. For larger levels of J_C, $\Delta\psi_C$ saturates at approximately $V_{CB} + 0.75\text{V}$. As understood from (2.118), this constancy of $\Delta\psi_C$ implies a linear relationship between $p_C(0)$ and J_C. The rate of increase in $p_C(0)$ with J_C is therefore greatly reduced by the quasisaturation effect, as illustrated in Figure 2.16.

The carrier concentration and field profiles in the collector are of particular interest in understanding the physics of the quasisaturation mode of operation. The hole concentration profile is determined by integrating (2.114). As depicted in Figure 2.17(a), $p_C(x_C)$ decreases linearly toward N_C in accordance with

$$p_C(x_C) = p_C(0) - \frac{q}{kT}\frac{J_C \mathscr{E}_o}{2qv_s}x_C = p_C(0) - \frac{J_C}{2qD_{nCo}}x_C$$

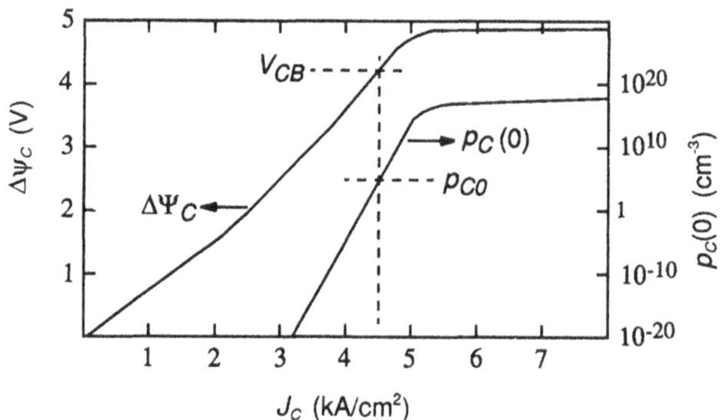

Figure 2.16 The effect of the collector current density J_C on the bulk voltage $\Delta\Psi_C$ and on the minority-carrier boundary concentration $p_C(0)$ in the active collector. The onset of the quasisaturation effect is marked with a vertical dashed line.

Once p_C falls sufficiently below N_C, the profile turns into an exponential one. Also shown in Figure 2.17(a) is the electron concentration profile $n_C(x_C)$, which is related to $p_C(x_C)$ through $n_C(x_C) = p_C(x_C) + N_C$ as required by quasineutrality. Obviously, the active collector region is divided into a high-injection zone, which is often referred to as the *extended base region*, and a low-injection zone known as the *ohmic region*.

The field profile can be readily obtained by substituting $p_C(x_C)$ into (2.113). The result can be approximated with

$$\mathscr{E}_C \cong -\frac{J_C \mathscr{E}_o}{2qv_s p_C} = -\frac{J_C}{2q\mu_{nCo} p_C}$$

for the extended base region, and by

$$\mathscr{E}_C \cong -\frac{J_C \mathscr{E}_o}{qv_s N_C} = -\frac{J_C}{q\mu_{nCo} N_C}$$

for the ohmic region. Since the $2p_C$ of the former is larger than the N_C of the latter, the field in the extended base is much weaker than the field in the ohmic region, as depicted in Figure 2.17(b). Also note that an increasing J_C moves the boundary between the two regions toward the buried layer.

Gummel Plot Under the Influence of the Webster and Quasisaturation Effects

The preceding analysis shows that, at the onset of quasisaturation, $\Delta\psi_C$ approximately equals $V_{CB} + 0.75\text{V}$ but $(kT/q)p_C(0)$ is still much smaller than $(\Delta\psi_C/2)N_C$. Relying on these observations, we can write from (2.118)

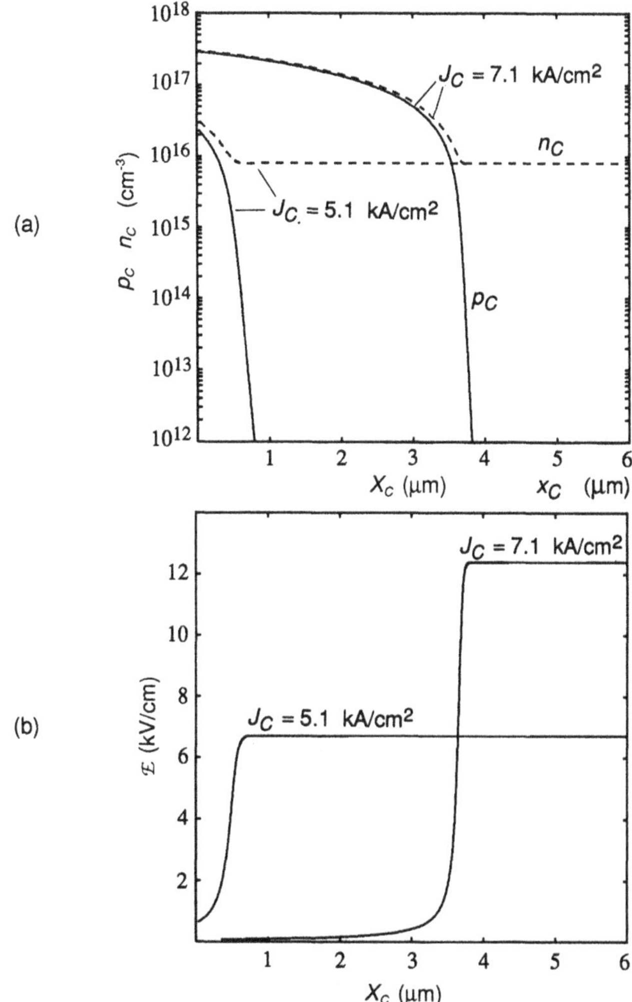

Figure 2.17 The development of an extended base in the collector region as a result of quasisaturation: (a) The region is characterized with high-level injection as indicated by the carrier concentrations being in excess of the doping concentration. (b) The extended base region is virtually field free, whereas the neighboring ohmic collector supports a large field, and therefore the entire collector bulk voltage. Also notice the expansion of the extended base with the increasing collector current density.

$$J_C = J_{C(\text{sat})} \equiv qv_sN_C\frac{V_{CB} + 0.75}{\mathscr{E}_o l_C + V_{CB} + 0.75}$$

which defines the boundary between the forward active and quasisaturation modes on the J_C-V_{CB} common-base output characteristics as shown in Figure 2.18. Recall

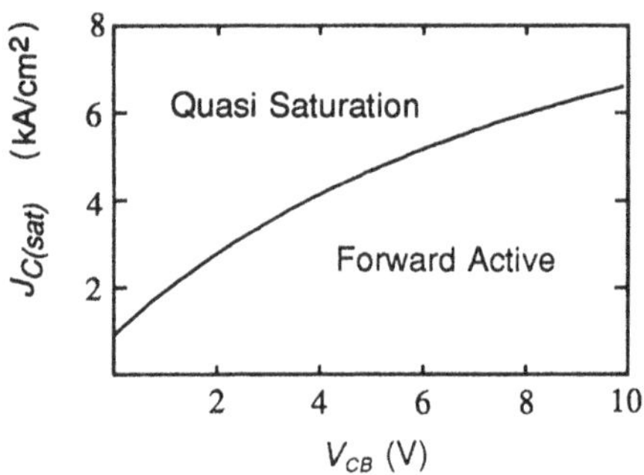

Figure 2.18 The locus of $J_C = J_{C(\mathrm{sat})}$, which demarcates the forward active and quasisaturation modes of operation.

that the basic model developed in Section 2.2 unconditionally predicts a forward active mode for all positive values of V_{CB}. Now we know that the forward active mode is confined to the low-current region below the locus of $J_C = J_{C(sat)}$. Now suppose that we keep V_{CB} constant but increase V_{BE} starting from this low-current region. We expect J_C to increase and eventually exceed $J_{C(sat)}$, forcing the device into the domain of the Webster and/or quasisaturation effects. We now discuss how the Gummel plot is affected by these high-bias phenomena.

In the general model presented in this section, the base and collector phenomena are represented by (2.103), (2.107) and (2.117), (2.118), respectively. A fifth independent equation, linking $n_B(W)$ to $p_C(0)$, can be obtained from the energy-band diagram of Figure 2.15(b). From a loop involving the quasi-Fermi potentials at the two boundaries of the CB transition region, we can write

$$\phi_{FnB}(W) - \phi_{FnC}(0) + \phi_{FpC}(0) - \phi_{FpB}(W) = 0$$

Rewriting this equation in terms of the carrier concentrations corresponding to these quasi-Fermi potentials, we obtain

$$\frac{n_B(W)}{n_C(0)} = \frac{p_C(0)}{p_B(W)}$$

Substituting into this equation $p_C(0) + N_C$ for $n_C(0)$ and $n_B(W) + N_B$ for $p_B(W)$, as required the condition of quasineutrality, we obtain the linking equation

$$n_B(W) = -\frac{N_B}{2} + \sqrt{\left(\frac{N_B}{2}\right)^2 + [p_C(0) + N_C]p_C(0)} \qquad (2.120)$$

By solving (2.103), (2.107), (2.117), (2.118), and (2.120) simultaneously, one can calculate J_C and the internal variables $\Delta\psi_C$, $p_C(0)$, $n_B(W)$, and $n_B(0)$ for a given set of structural parameters plus V_{BE} and V_{CB}. Solutions for a typical BJT of $N_B = 10^{17}$ cm^{-3}, $W = 1$ µm, $N_C = 8 \times 10^{15}$ cm^{-3}, and $l_C = 10$ µm are given in Figure 2.19 as functions of V_{BE} for $V_{CB} = 4$V.

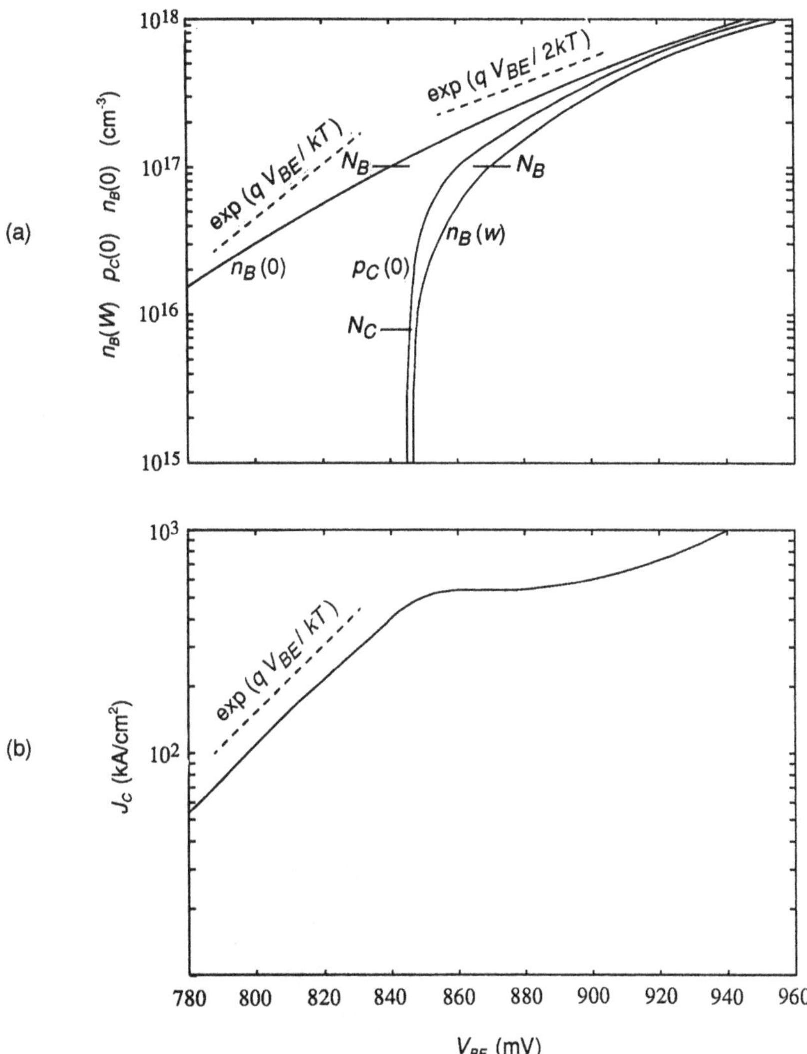

Figure 2.19 (a) The variation of minority-carrier boundary concentrations in the base and the collector with the base-emitter voltage. The intersection of $n_B(0)$ with N_B marks the onset of high-level injection at the emitter side of the base (Webster effect). The base falls under high-level injection conditions in its entirety when $n_B(W)$ also exceeds N_B. High-level injection in the collector is marked by the intersection of $p_C(0)$ with N_C. (b) Collector current Gummel plot under the influence of the Webster and quasisaturation effects.

The device enters quasisaturation at about $V_{BE} = 0.85$V. Notice from the $n_B(0)$ versus V_{BE} plot that the device is already under the influence of a moderate Webster effect when quasisaturation occurs. This effect reaches full strength for $V_{BE} > 900$ mV, as observed from the halved slope of the same plot.

The slope of the J_C Gummel plot also decreases gradually under the influence of the Webster effect before the occurrence of quasisaturation. Thereafter, the rate declines sharply. For an explanation of this sharp decline, first recall the conclusion of (2.103) that J_C is proportional to the difference $n_B(0) - n_B(W)$ regardless of the level of bias, that is

$$J_C = m\frac{qD_{nB}}{W}[n_B(0) - n_B(W)] \tag{2.121}$$

where $1 \le m \le 2$ accommodates the rate enhancement due to the Webster effect. Next, note from (2.120) that $n_B(W)$ is an increasing function of $p_C(0)$. Assuming $N_C \ll p_C(0) \ll N_B/2$ for the relatively mild initial phases of quasisaturation following the onset, the variation of $n_B(W)$ with $p_C(0)$ can be determined from the following series approximation of (2.120):

$$n_B(W) \cong \frac{p_C^2(0)}{N_B} \tag{2.122}$$

On the other hand, $p_C(0)$ becomes a linearly increasing function of J_C in quasisaturation, as (2.118) indicates for a constant $\Delta\psi_C$. The rate can be obtained from (2.118) as

$$\frac{dp_C(0)}{dJ_C} = \frac{q}{kT}\,\frac{\mathscr{E}_o l_C + V_{CB} + 0.75}{2qv_s} \tag{2.123}$$

Turning back to (2.121), we can express the rate of the Gummel plot as

$$\frac{dJ_C}{J_C dV_{BE}} = m\frac{qD_{nB}}{WJ_C}\left[\frac{dn_B(0)}{dV_{BE}} - \frac{dn_B(W)}{dp_C(0)}\,\frac{dp_C(0)}{dJ_C}\,\frac{dJ_C}{dV_{BE}}\right]$$

Substituting (2.123) and the derivative of (2.122) into this equation, using (2.121), rearranging, and finally assuming $n_B(W) \ll n_B(0)$ for the mild quasisaturation, we obtain

$$\frac{dJ_C}{J_C dV_{BE}} = \frac{1}{1 + [m\mu_{nB}(\mathscr{E}_o l_C + V_{CB} + 0.75)p_C(0)/WN_B v_s}\,\frac{dn_B(0)}{n_B(0)\,dV_{BE}} \tag{2.124}$$

We know from the two asymptotic forms of (2.107) that the rate $dn_B(0)/n_B(0)\,dV_{BE}$ changes from q/kT in low bias to $q/2kT$ in high bias under the influence of the

Webster effect. The pre-quasisaturation slope of the Gummel plot is determined solely by this rate. Its prefactor in (2.124) represents the effect of quasisaturation. This, being inversely proportional to $p_C(0)$, rapidly decreases below unity with the increasing bias, thus causing the abrupt decline in the slope of the Gummel plot. However, we observe a recovery of the slope at higher bias levels. According to the present model, this happens as the collector side of the base undergoes high-level injection conditions. In the extreme, (2.120) indicates $n_B(W) \cong p_C(0)$, hence a weakening of the dependence of $n_B(W)$ on $p_C(0)$ in comparison with (2.122). This is why the slope recovers in strong quasisaturation.

The base current is virtually unaffected by the Webster and quasisaturation effects, while these effects make the collector current less sensitive to V_{BE}. Therefore, the apparent value of β_F is expected to decline at high levels of bias, as shown in Figure 2.20. In practice, this high-bias deterioration of β_F is indeed observed in all BJTs, and in most cases, it is accompanied by other high-bias effects, such as finite base resistance and current crowding. For this reason, the high-bias behavior in quantitative terms is usually more complicated than that predicted by the Webster and quasisaturation models alone. This is also why a large number of supplementary models and alternative descriptions have been offered for high-bias effects in BJTs. One such description, the so-called *Kirk effect*, is frequently cited as the dominant high-bias effect. A quantitative discussion of this effect is presented next.

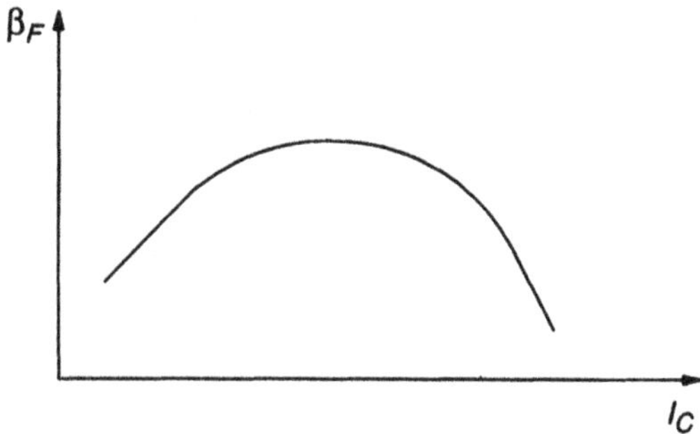

Figure 2.20 A typical variation of β_F with I_C for all bias levels.

Kirk Effect

In remodeling the collector for high bias, we implicitly assumed a fixed active collector length l_C. Although the location of the common boundary with the buried layer is fixed metallurgically, the boundary with the CB transition region can be displaced not only by V_{CB} under the influence of the Early effect, but also by the collector current. Indeed, an increasing J_C may expand the CB transition region further into the collector, unless a simultaneous increase in $\Delta\psi_C$ adequately debiases the transition region. To understand why, consider the fact that an increasing J_C

must be accompanied with an increasing concentration of electrons transiting the CB transition region. Once this concentration becomes comparable to N_C, the density of the space charge on the collector side of the transition region is described by $\rho = q(N_C - n)$. Using this equation instead of $\rho = qN_C$ in the derivation of (2.83), we obtain the width of the space-charge region on the collector side as

$$x_2 = \sqrt{\frac{2\epsilon}{q(N_C - \bar{n})}[V_{CB} - \Delta\psi_C + V_{b(CB)}]}$$

where $\bar{n}$ denotes the average concentration of electrons in the transition region. Also note that the term V_{CB} of (2.83) is replaced by $V_{CB} - \Delta\psi_C$, which, of course, is the proper expression of the transition region bias in the case of nonnegligible collector bulk voltage $\Delta\psi_C$. Since both $\bar{n}$ and $\Delta\psi_C$ are increasing functions of J_C, x_2 may increase or decrease with J_C depending the structural features of the collector and the bias. A decreasing width has little, if any, effect on a quasisaturation-dominated high-bias behavior. An increasing width, on the other hand, reduces l_C, and hence the collector resistance R_C, thus moderating the quasisaturation.

According to a widely recognized model, initially offered by Kirk in 1961, once the space-charge region expands all the way to the high-low junction, any further increase in J_C for a constant V_{CB} should result in a passage of the base-side boundary of the same region into the collector. [16]. In other words, the base is pushed out into the collector with a net effect of increasing width W. A reduction of the sensitivity of J_C to V_{BE} is thus implied. This so-called *Kirk effect* is believed to replace quasisaturation for relatively large values of V_{CB}. The reader is referred to Problem 2.6 for an analysis of the Kirk effect.

2.3.3 Impact Ionization, Avalanche Multiplication, and Junction Breakdown

Physical Origins

The impact ionization and avalanche multiplication effects and their ultimate manifestation—junction breakdown—can occur in reverse-biased junctions, [17], for example the CB junction in the forward active mode of operation [18]. For a qualitative description of these effects, consider the reverse-biased CB junction schematic of Figure 2.21. Holes and electrons are seen to diffuse into the transition region from the bulk regions where they are minority carriers. These carriers join the ones generated thermally in the transition region, where they all drift into the bulk regions and become majority carriers. As we increase V_{CB}, the field in the transition region increases, which, in turn, enhances the velocity and therefore the energy of the carriers in transition. Any carrier, having acquired a sufficient amount of energy to ionize a silicon atom upon collision, can now create an additional

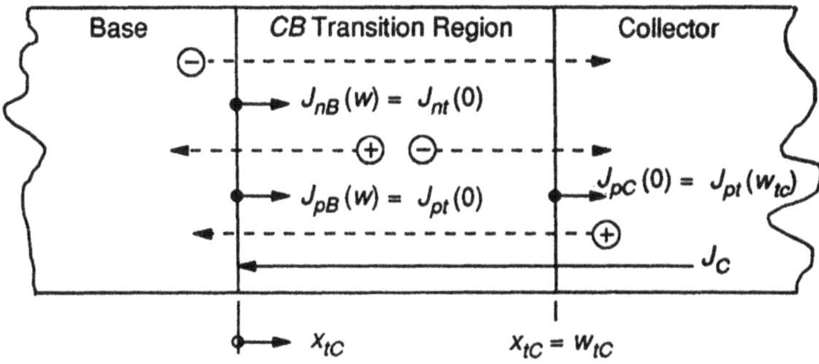

Figure 2.21 The variables involved in the analysis of impact ionization, avalanche multiplication, and avalanche breakdown.

electron-hole pair before leaving the transition region. This is what we call *impact ionization*. The pair thus created splits under the influence of the electric field. The electron drifts toward the collector while the hole drifts in the opposite direction. If the field is high, these additional carriers, too, can acquire sufficient energy to trigger new impact ionization events before they exit the transition region. This leads to an *avalanche multiplication* of carrier concentrations in the region. In consequence, the collector current becomes an increasing function of V_{CB}. Ultimately, the rate of change of I_C with V_{CB} reaches infinity, which marks the *breakdown* of the CB junction.

A Mathematical Model for Avalanche Multiplication

For a quantitative modeling of the avalanche multiplication effect, we first rewrite the hole continuity equation (1.25) for the CB transition region in steady state without ignoring g. The latter now represents the number of holes generated by impact ionization per unit time per unit volume of the CB transition region:

$$\frac{dJ_{pt}}{dx_{tC}} = q(G_{th} - R) + qg \tag{2.125}$$

Now we formulate g in terms of impact ionization parameters and collector current density. Suppose that an electron ionizes αdx_{tC} carrier pairs as it traverses a distance dx_{tC} within the transition region. If it takes a time dt to traverse this distance, the number of pairs created by a single electron in unit time will be $\alpha dx_{tC}/dt = \alpha v_n$, where v_n is the magnitude of the electron velocity. Considering that there are n_t electrons in the unit volume, the total number of pairs generated by electrons per unit volume per unit time will be $\alpha n_t v_n$. The total number of pairs generated by holes per unit volume per unit time can be expressed similarly: $\alpha p_t v_p$.[3] Obviously, g equals the sum of these two, that is,

$$g = \alpha(n_t v_n + p_t v_p) \tag{2.126}$$

where α is the *ionization coefficient*. Carrier velocities are related to the drift-dominated currents through $v_n = -J_{nt}/qn_t = (J_C + J_{pt})/qn_t$ and $v_p = -J_{pt}/qp_t$.[4] Rewriting (2.126) with these equations, and substituting the outcome for g in (2.125), we obtain

$$\frac{dJ_{pt}}{dx_{tC}} = q(G_{th} - R) + \alpha J_C$$

whose integration over the transition region yields

$$J_{pt}(W_{tC}) = J_{pt}(0) + q\int_0^{W_{tC}} (G_{th} - R)\, dx_{tC} + J_C\int_0^{W_{tC}} \alpha\, dx_{tC}$$

We can also write from Figure 2.21 the equation $J_{pt}(W_{tC}) = J_{pC}(0)$ and $J_{pt}(0) = J_{pB}(W) = -J_C - J_{nB}(W)$. Substituting these together with (2.17) into the previous equation and rearranging, we obtain

$$J_C = M[-J_{pC}(0) - J_{nB}(W) + J_{grC}] \tag{2.127}$$

where M is the *multiplication factor* and is defined by

$$M \equiv \frac{1}{1 - \int_0^{W_{tC}} \alpha\, dx_{tC}} \tag{2.128}$$

Equation (2.127) is a more general form of (2.16). It shows that all of the diffusion and thermal-generation components of the collector current are enhanced by the multiplication factor as a result of impact ionization events. Equations (2.32), (2.28), and (2.43) still describe $J_{pC}(0)$, $J_{nB}(W)$, and J_{grC}, respectively, because neither the minority-carrier currents nor the thermal generation process is affected

[3]The value of α for holes is different from its counterpart for electrons. For the sake of simplicity, however, we assume they are equal in this analysis.

[4]We introduce these minus signs because (1) by definition, v_n and v_p are the absolute values of carrier velocities and (2) the actual directions of J_{nt} and J_{pt} are opposite to the reference directions shown in Figure 2.21.

by impact ionization. It is important to know that the ionization coefficient α is an increasing function of the electric field. An increasing V_{CB} enhances α by increasing the field in the transition region. As a result, the so-called *ionization integral* appearing in the denominator of (2.128) approaches unity, and therefore M starts to increase. This is why I_C is an increasing function of V_{CB} under the influence of the avalanche multiplication effect. It is left to the reader to show that the normally positive base current becomes a decreasing function of V_{CB} due to the same effect, and that even a negative base current is possible in the case of excessive multiplication.

The Effect on the Ebers-Moll Equations

The effect of avalanche multiplication on the Ebers-Moll model can be represented by multiplying the collector current equation (2.45) by M. The emitter current equation (2.44) is unaffected because none of the internal components of I_E is subject to multiplication. Considering the fact that $V_{CB} \gg kT/q$ for the multiplication to occur, these equations can be rewritten as

$$I_E = -\frac{I_s}{\alpha_F}\left[\exp\left(\frac{q}{kT}V_{BE}\right) - 1\right] - I_S \tag{2.129}$$

$$I_C = MI_S\left[\exp\left(\frac{q}{kT}V_{BE}\right) - 1\right] + M\frac{I_S}{\alpha_R} \tag{2.130}$$

whereas the base current equation becomes

$$I_B = \frac{I_S}{\alpha_F}(1 - M\alpha_F)\left[\exp\left(\frac{q}{kT}V_{BE}\right) - 1\right] + \frac{I_S}{\alpha_R}(\alpha_R - M) \tag{2.131}$$

Canceling V_{BE} between (2.130) and (2.129) yields the following equation of the output characteristics in the common-base configuration:

$$I_C = -M\alpha_F I_E + \frac{MI_S}{\alpha_R}(1 - \alpha_F\alpha_R) \tag{2.132}$$

Similarly, we can obtain the common-emitter output characteristics by canceling V_{BE} between (2.130) and (2.131):

$$I_C = \frac{M\alpha_F}{1 - M\alpha_F}I_B + \frac{MI_S}{\alpha_R}\frac{1 - \alpha_F\alpha_R}{1 - M\alpha_F} \tag{2.133}$$

Avalanche Breakdown

Now consider *junction breakdown*, which, as mentioned previously, is the ultimate form of avalanche multiplication causing $I_C \to \infty$. In the common-base configuration, the breakdown is characterized by $BV_{CBO} \equiv V_{CB}|_{I_E=0, I_C \to \infty}$. It is obvious from (2.132) that if avalanche breakdown is to occur, M must approach infinity. As a matter of fact, the most popular empirical equation describing the relationship between M and V_{CB} is expressed in terms of BV_{CBO} as follows:

$$M = \frac{1}{1 - (V_{CB}/BV_{CBO})^n} \tag{2.134}$$

where n is typically 3.

In the common-emitter configuration, the breakdown voltage is defined as $BV_{CEO} \equiv V_{CE}|_{I_B=0, I_C \to \infty}$. According to (2.133), breakdown occurs for $M = 1/\alpha_F \cong 1$ instead of $M \to \infty$. Replacing M with $1/\alpha_F$ and V_{CB} with BV_{CEO} (because V_{BE} is small), and solving (2.134) for BV_{CEO}, we obtain

$$BV_{CEO} \cong BV_{CBO}(1 - \alpha_F)^{1/n} \cong \frac{BV_{CBO}}{\beta_F^{1/n}} \tag{2.135}$$

which indicates that BV_{CEO} is considerably smaller than BV_{CBO}.

The breakdown voltage depends heavily on the structural parameters such as the width and doping concentration of the active collector region and the curvature of the CB junction along its periphery. In a lightly doped active collector, the electric field is a weak function of V_{CB}, which implies a higher breakdown voltage. If, however, the buried layer is close to the junction, so that the entire active collector is depleted before breakdown occurs, any further increase in V_{CB} will rapidly enhance the field and thus cause a smaller-than-expected breakdown voltage, which is called *punch-through-limited* breakdown. The field is enhanced also along the periphery of the junction due to the finite radius of curvature, which also reduces the breakdown voltage. These effects are illustrated in Figure 2.22 with the BV_{CEO} curves plotted as functions of β_F, N_C, W_C, and junction curvature [19].

2.4 BJT DYNAMICS

In this section we model the time-dependent BJT behavior. First, consider a BJT that has been operating in the forward active mode under dc steady-state conditions, and suppose that the base-emitter voltage is increased instantaneously at time $t = T$. As shown in Figure 2.23(a), carrier concentrations at the EB transition region boundaries will increase rapidly in accordance with (2.21) and (2.26) resulting in enhanced carrier transport into the neighboring bulk regions. Since the

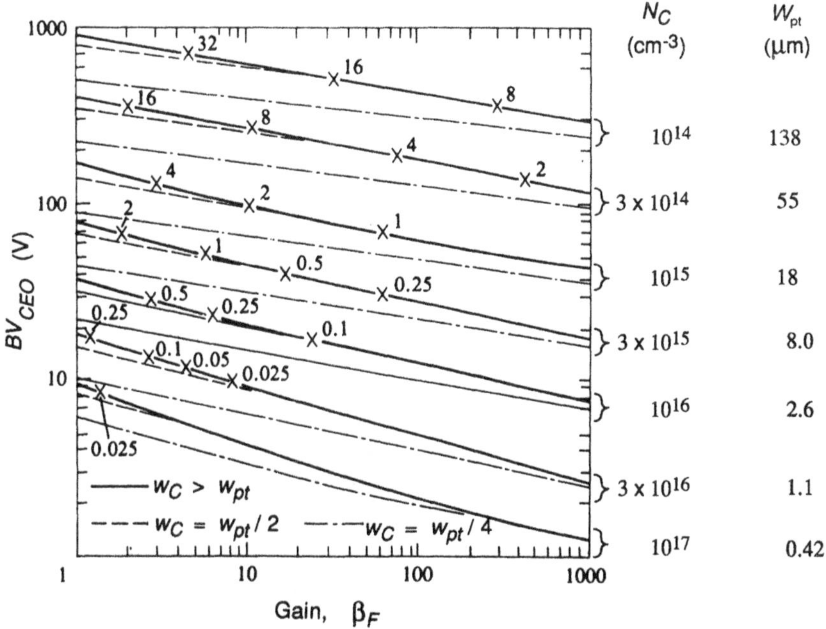

Figure 2.22 The BV_{CEO} curves as a function of β_F, N_C, W_C, and junction curvature. The solid curves belong to long-collector BJTs, in which, breakdown occurs before the CB transition region reaches the buried layer. If the active collector length W_C is shorter than a critical length W_{pt}, the breakdown voltage is reduced by punch-through, as indicated by the curves representing the cases of $W_C = W_{pt}/2$ and $W_C = W_{pt}/4$. The numbers marked on each curve represent, in micrometer units, the minimum radius of junction curvature necessary for realizing the breakdown voltage indicated by the curve. (From Roulston and Depey [19]. © 1980 IEE. Reprinted with the permission of the publisher.)

transport process advances at a finite speed, it takes a certain time for the bulk carrier profiles to reach their new steady-state forms, as shown in the same figure. These transient transport processes can be characterized by solving the partial differential equations obtained from current density equations and time-dependent continuity equations. For example, in the emitter and base, where generation and recombination effects are negligible and minority carriers move by diffusion only, (1.18), (1.24) and (1.13), (1.23) yield the following differential equations:

$$D_{nB}\frac{d^2n_B(x,t)}{dx^2} = \frac{dn_B(x,t)}{dt}$$

$$D_{pE}\frac{d^2p_E(x,t)}{dx^2} = \frac{dp_E(x,t)}{dt}$$

which must be solved with appropriate boundary and initial conditions. However, the complicated nature of these partial differential equations severely limits their

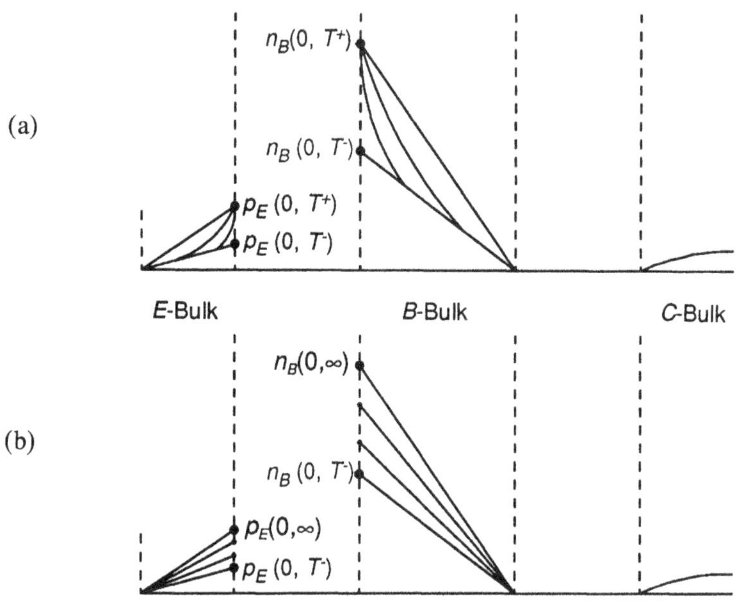

Figure 2.23 Evolution of the minority-carrier concentration profiles in response to an increasing base-emitter voltage in the forward active mode: (a) the case of a rapidly varying stimulus and (b) the case of a slowly varying stimulus.

usefulness in an engineering model. Fortunately, the time-dependent excitation of a BJT rarely involves rates that exceed the speed of the internal transport processes mentioned earlier. Illustrated in Figure 2.23(b) is the transient process of carrier profile evolution in response to a gradually increasing V_{BE}. Notice that, since the carrier boundary concentrations $n_B(0,t)$ and $p_E(0,t)$ increase gradually, the profiles are not distorted by the time-dependent transport processes; they approximately retain their steady-state forms at all times during the transition. This is very fortunate because we can still use the equations previously derived for the steady-state forms of these profiles, such as (2.19), (2.25), and (2.30). This assumption of *quasistatic* response leads to what is known as the *charge-control model* of the BJT. We now discuss the derivation of this model and its representation with equivalent circuits. For further reading on the charge-control formulation, the reader is referred to Warner and Grung [20]. A derivative, called the *Gummel-Poon model* is of great utility in numerical simulation of BJTs. A detailed presentation of this model can be found in Antognetti and Massobrio [21].

2.4.1 The Quasistatic BJT Behavior

Charge-Control Modeling of the Emitter

Figure 2.24 gives the reference convention of the charge-control model. We begin by integrating the time-dependent electron continuity equation (1.24) over position between $x = 0$ and $x = x_1$ assuming negligible generation and recombination in the emitter bulk and in the EB transition region. Also ignoring the time variation of the small electron density per unit area in the EB transition region, we obtain

$$i_n(x_1) - i_n(0) = \frac{d}{dt}\int_0^{x_1'} Aqn_E \, dx \tag{2.136}$$

Note equation $i_n(0) = i_E - i_p(0)$ from Figure 2.24, and $n_E = N_E + n_E' \cong N_E + p_E'$ from the condition of quasineutrality in the emitter bulk. With these, we can rewrite (2.136) as

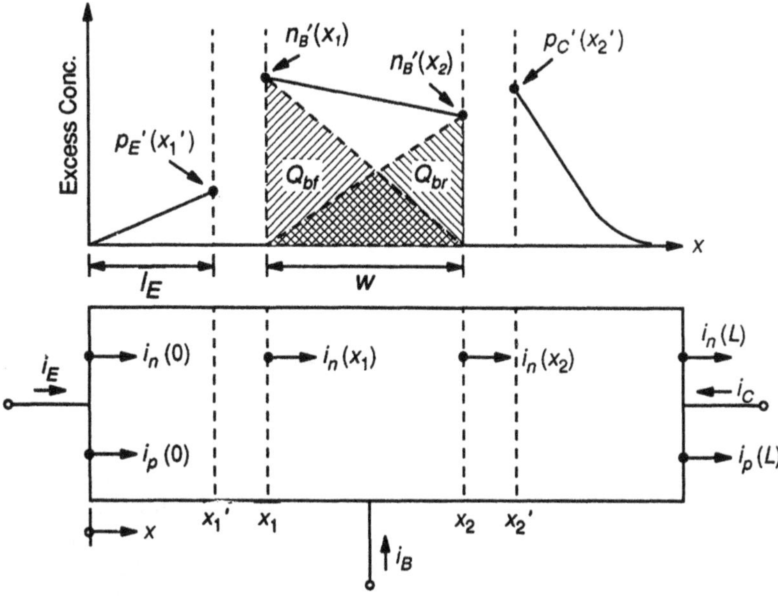

Figure 2.24 The time-dependent variables and the reference convention involved in the derivation of the charge-control model.

$$i_E = i_n(x_1) + i_p(0) - \frac{d}{dt}\int_0^{x_1'} AqN_E\,dx - \frac{dQ_e}{dt} \tag{2.137}$$

where

$$Q_e \equiv Aq\int_0^{x_1'} p_E'\,dx \tag{2.138}$$

is the total charge of excess holes in the n-type emitter bulk.

Now consider the hole current in the emitter, $i_p(0)$. Being a bulk minority current, it flows by diffusion, that is, $i_p(0) = -AqD_{pE}(dp_E'/dx)$. As shown in Figure 2.24, p_E' retains its linear profile as a result of the quasistatic operation. Therefore,

$$i_p(0) = -A\frac{qD_{pE}}{x_1'}p_E'(x_1') \tag{2.139}$$

Using this linear profile in (2.138), we obtain

$$Q_e = A\frac{qx_1'}{2}p_E'(x_1') \tag{2.140}$$

Now, canceling $p_E'(x_1')$ between (2.139) and (2.140), we arrive at

$$i_p(0) = -\frac{Q_e}{\tau_e} \tag{2.141}$$

where

$$\tau_e \equiv \frac{(x_1')^2}{2D_{pE}} = \frac{l_E^2}{2D_{pE}} \tag{2.142}$$

is a time constant associated with the emitter bulk. Soon, we will substitute (2.141) for $i_p(0)$ in (2.137). But first, consider the third term on the right-hand side of (2.137). Recalling definition (2.73) of the emitter Gummel number, we can express that term as $Aq(dG_e'/dt)$ where G_e' can vary with time because the boundary x_1' is a function of the generally time-dependent v_{EB} due to an Early-like effect associated with the EB junction. But since $G_e' = G_{em} - G_{eb}$ where G_{em} is the voltage-independent metallurgical emitter Gummel number and G_{eb} is the net doping concentration per unit area on the emitter side of the EB transition region, dG_e'/dv_{EB} must equal $-dG_{eb}/dv_{EB}$. Moreover, $G_{eb} = G_{be}$ because the charges on the two sides of a pn junction transition region have the same magnitude. Therefore, the third term on the right-hand side of (2.137) can be expressed as $-C_{je}\, dv_{EB}/dt$, where C_{je} is defined by

$$C_{je} \equiv Aq\frac{dG_{be}}{dv_{EB}} \tag{2.143}$$

and is called *EB junction capacitance*. Using this together with (2.141) in (2.137), we finally obtain

$$i_E = i_n(x_1) - \frac{Q_e}{\tau_e} + C_{je}\frac{dv_{EB}}{dt} - \frac{dQ_e}{dt} \tag{2.144}$$

Charge-Control Modeling of the Collector

Now we integrate the time-dependent electron continuity equation (1.24) over position between $x = x_2$ and $x = l$. Assuming negligible generation and recombination in the CB transition region, and ignoring the time variation of the small electron density per unit area in this region, we find

$$i_n(l) - i_n(x_2) = \frac{d}{dt}\int_{x_2'}^{l} Aqn_C\, dx - Aq\int_{x_2'}^{l}(G_{th} - R)\, dx \tag{2.145}$$

Note that $G_{th} - R$ in the collector is proportional to $p_C'(x)$ in accordance with (1.78). Based on the assumption of quasistatic operation, we still expect the profile of $p_c'(x)$ to obey (2.30) that is,

$$p_C'(x) = p_C'(x_2')\exp\left(-\frac{x - x_2'}{L_{pC}}\right) \tag{2.146}$$

Since L_{pC} is assumed to be much smaller than the active collector length, we also expect the diffusion-based $i_p(x)$ to vanish well before the boundary $x = l$ is reached. Therefore $i_p(l) = 0$, and thus $i_C = -i_n(l)$. Taking these into account, and adopting the mathematical procedure previously applied to (2.136), we finally obtain the following equation from (2.145):

$$i_C = -i_n(x_2) - \frac{Q_c}{\tau_c} + C_{jc}\frac{dv_{CB}}{dt} - \frac{dQ_c}{dt} \tag{2.147}$$

where

$$Q_c \equiv Aq\int_{x_2'}^{l} p_C'\, dx = AqL_{pC}p_C'(x_2') \tag{2.148}$$

is the total charge of excess holes in the n-type collector bulk,

$$\tau_c = \tau_{pC} \tag{2.149}$$

is a time constant associated with the collector bulk, and

$$C_{jc} \equiv -Aq\frac{d}{dv_{CB}}\left(\int_{x_2'}^{l} N_C \, dx\right) = Aq\frac{dG_{bc}}{dv_{CB}} \tag{2.150}$$

is the CB junction capacitance.[5]

Charge-Control Modeling of the Base

To complete the model, we must also characterize $i_n(x_1)$ and $i_n(x_2)$, which appear in (2.144) and (2.147). For this purpose, we integrate the time-dependent electron continuity equation (1.24) over position between $x = x_1$ and $x = x_2$. Assuming negligible generation and recombination in the base, we obtain

$$i_n(x_2) - i_n(x_1) = \frac{dQ_b}{dt} \tag{2.151}$$

where

$$Q_b \equiv Aq\int_{x_1}^{x_2} n_B' \, dx \tag{2.152}$$

is the total charge of excess electrons in the p-type base. Due to the assumption of quasistatic operation, we still expect the linear $n_B'(x)$ profile to prevail as shown in Figure 2.24. This enables us to divide Q_b into two parts:

$$Q_b = Q_{bf} + Q_{br} \tag{2.153}$$

where

$$Q_{bf} \equiv Aq\frac{W}{2}n_B'(x_1) \tag{2.154}$$

and

$$Q_{br} \equiv Aq\frac{W}{2}n_B'(x_2) \tag{2.155}$$

[5]Note that the derivation of this expression of C_{jc} is similar to the derivation of (2.143).

The shaded triangular areas in Figure 2.24 represent these *forward* and *reverse* base charges. Now, consider the dc steady-state values $I_n(x_1)$ and $I_n(x_2)$ of the boundary currents. From (2.23), (2.24), (2.25), (2.154), and (2.155) we can easily obtain

$$I_n(x_1) = I_n(x_2) = -\frac{Q_{bf} - Q_{br}}{\tau_b} \tag{2.156}$$

where

$$\tau_b = \frac{W^2}{2D_{nB}} \tag{2.157}$$

is called the *base transit time*. Now consider the general transient state in which Q_{bf} and Q_{br} change with time. Since Q_{bf} is proportional to $n'_B(x_1)$, any change in the former must be accompanied by a change in the latter. As the electron supplying current at $x = x_1$, $i_n(x_1)$, therefore, can be assumed solely responsible for the time-dependent variation of Q_{bf}. This enables us to write

$$i_n(x_1) = I_n(x_1) - \frac{dQ_{bf}}{dt} = -\frac{Q_{bf}}{\tau_b} + \frac{Q_{br}}{\tau_b} - \frac{dQ_{bf}}{dt} \tag{2.158}$$

Similarly, we expect $i_n(x_2)$ to be solely responsible for any time-dependent variation of Q_{br}, hence:

$$i_n(x_2) = -\frac{Q_{bf}}{\tau_b} + \frac{Q_{br}}{\tau_b} + \frac{dQ_{br}}{dt} \tag{2.159}$$

Note that (2.151) is indeed satisfied by (2.158) and (2.159).

Charge-Control Model Equations

Using (2.158) and (2.159) in (2.144) and (2.147), we obtain the following charge-control equations describing the quasistatic BJT operation:

$$i_E = \frac{Q_{br}}{\tau_b} - \left(\frac{Q_{bf}}{\tau_b} + \frac{Q_e}{\tau_e}\right) + C_{je}\frac{dv_{EB}}{dt} - \frac{d}{dt}(Q_{bf} + Q_e) \tag{2.160}$$

$$i_C = \frac{Q_{bf}}{\tau_b} - \left(\frac{Q_{br}}{\tau_b} + \frac{Q_c}{\tau_c}\right) + C_{jc}\frac{dv_{CB}}{dt} - \frac{d}{dt}(Q_{br} + Q_c) \tag{2.161}$$

These charge-control equations can also be expressed as follows:

$$i_E = \frac{Q_R}{\tau_R} - \frac{1 + \beta_F}{\beta_F}\frac{Q_F}{\tau_F} + C_{je}\frac{dv_{EB}}{dt} - \frac{dQ_F}{dt} \tag{2.162}$$

$$i_C = \frac{Q_F}{\tau_F} - \frac{1 + \beta_R}{\beta_R}\frac{Q_R}{\tau_R} + C_{jc}\frac{dv_{CB}}{dt} - \frac{dQ_R}{dt} \tag{2.163}$$

where

$$Q_F \equiv Q_{bf} + Q_e \tag{2.164}$$

$$Q_R \equiv Q_{br} + Q_c \tag{2.165}$$

$$\tau_F \equiv \frac{Q_F}{Q_{bf}}\tau_b \tag{2.166}$$

$$\tau_R \equiv \frac{Q_R}{Q_{br}}\tau_b \tag{2.167}$$

where τ_F and τ_R are called *forward transit time* and *reverse transit time*, respectively. Equations (2.162) and (2.163) describe i_E and i_C in terms of two specific charges, Q_F and Q_R. The former charge is controlled exclusively by v_{EB}, the latter by v_{CB}. This enables us to express dQ_F/dt and dQ_R/dt as

$$\frac{dQ_F}{dt} = -C_{de}\frac{dv_{EB}}{dt} \tag{2.168}$$

$$\frac{dQ_R}{dt} = -C_{dc}\frac{dv_{CB}}{dt} \tag{2.169}$$

where

$$C_{de} \equiv -\frac{dQ_F}{dv_{EB}} \tag{2.170}$$

$$C_{dc} \equiv -\frac{dQ_R}{dv_{CB}} \tag{2.171}$$

are known as *diffusion capacitances*. With these definitions, the charge-control relationships are transformed into the following form:

$$i_E = I_E + (C_{je} + C_{de})\frac{dv_{EB}}{dt} \tag{2.172}$$

where

$$I_E = -\left(\frac{Q_F}{\tau_F} - \frac{Q_R}{\tau_R}\right) - \frac{Q_F}{\beta_F\tau_F}$$

and

$$i_C = I_C + (C_{jc} + C_{dc})\frac{dv_{CB}}{dt} \tag{2.173}$$

where

$$I_C = \frac{Q_F}{\tau_F} - \frac{Q_R}{\tau_R} - \frac{Q_R}{\beta_R\tau_R}$$

The charge-control relationship for the base current can be constructed from (2.172), (2.173), and $i_B = -i_E - i_C$ as follows:

$$i_B = I_B - (C_{je} + C_{de})\frac{dv_{EB}}{dt} - (C_{jc} + C_{dc})\frac{dv_{CB}}{dt} \tag{2.174}$$

where

$$I_B = \frac{Q_F}{\beta_F\tau_F} + \frac{Q_R}{\beta_R\tau_R}$$

2.4.2 BJT Equivalent Circuits

Large-Signal Equivalent Circuits

Two alternative *large-signal equivalent circuits* representing the charge-control model of a BJT are given in Figure 2.25. Also included in these equivalent circuits are the resistances R_B, R_C, and R_E of the quasineutral bulk regions and the junction capacitance C_{cs} of the parasitic collector-substrate junction. Excluding the parasitic components, the reader can easily verify the correspondence between these equivalent circuits and the charge-control model equations (2.172), (2.173), and (2.174).

Note that all five capacitances appearing in the equivalent circuits are bias dependent. The bias dependence of the diffusion capacitances can be described with the aid of (2.170), (2.171), (2.164) through (2.167), and (2.47) as follows.

$$C_{de} = \tau_F I_S \frac{q}{kT} \exp\left(-\frac{q}{kT}v_{EB}\right) \tag{2.175}$$

$$C_{dc} = \tau_R I_S \frac{q}{kT} \exp\left(-\frac{q}{kT}v_{CB}\right) \tag{2.176}$$

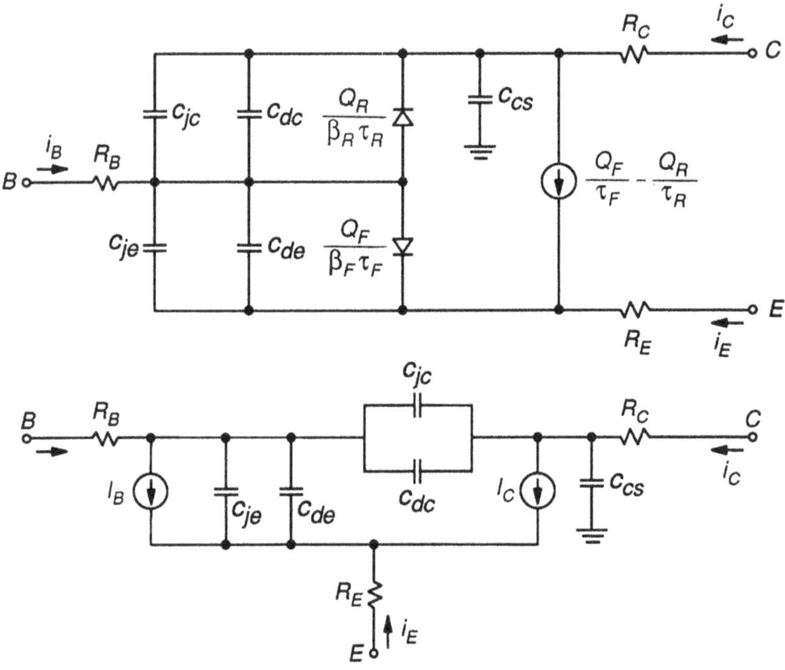

Figure 2.25 The large-signal equivalent circuits representing the charge-control model of a BJT.

The bias dependence of the junction capacitances can be modeled on the basis of the analysis presented in Section 2.2.7 for the Early effect. Assuming depletion in the transition region, the bias dependence is found to have the form

$$C_j = \frac{C_j(0)}{[1 + (v/V_b)]^m} \tag{2.177}$$

for a reverse-biased or mildly forward-biased junction regardless of the doping profile gradient. Here, $C_j(0)$ is the zero-bias capacitance, v is the bias voltage (v_{EB} or v_{CB}), V_b is the built-in voltage, and m is the so-called *gradient factor* whose value is 0.5 for an *abrupt junction* (appropriate for the CB junction) and 0.33 for a *linearly graded junction* (appropriate for the EB junction). Note that the collector-substrate capacitance C_{cs} can also be modeled with (2.177).

A Small-Signal Equivalent Circuit

The equivalent circuits of Figure 2.25 can be linearized for those cases in which the BJT operates in the forward active mode with an ac base-emitter voltage of small amplitude. Most analog circuit applications fall into this category. To derive this equivalent circuit, we will express the quasistatic components I_B and I_C of (2.174) and (2.173) as

$$I_B \Rightarrow \frac{I_S(0)}{\beta_F(0)} \exp\left(\frac{q}{kT} v_{BE}\right) \tag{2.178}$$

$$I_C \Rightarrow I_S(0)\left(1 + \frac{v_{CB}}{V_{AF}}\right) \exp\left(\frac{q}{kT} v_{BE}\right) \tag{2.179}$$

where $I_S(0)$ and $\beta_F(0)$ represent I_S and β_F, respectively, for $V_{CB} = 0$. Note that a factor $1 + (v_{CB}/V_{AF})$ is included in (2.179) to account for the Early effect. It can be easily derived by recognizing $I_C/I_C(0) = I_S/I_S(0) = 1 + (V_{CB}/V_{AF})$ from Figure 2.10. By using (2.178) and (2.179), we can transform (2.173) and (2.174) into

$$i_C = I_S(0)\left(1 + \frac{v_{CB}}{V_{AF}}\right) \exp\left(\frac{q}{kT} v_{BE}\right) + (C_{jc} + C_{dc})\frac{dv_{CB}}{dt}$$

$$i_B = \frac{I_S(0)}{\beta_F(0)} \exp\left(\frac{q}{kT} v_{BE}\right) + (C_{je} + C_{de})\frac{dv_{BE}}{dt} - (C_{jc} + C_{dc})\frac{dv_{CB}}{dt}$$

Expanding these functions into a series around the dc bias values, and ignoring the second-order terms, we obtain the following small-signal ac characteristic equations[6]:

$$I_c = \frac{1}{r_e} V_{be} + \frac{1}{r_o} V_{ce} + j\omega C_\mu V_{cb} \tag{2.180}$$

$$I_b = \frac{1}{\beta_F r_e} V_{be} + j\omega C_\pi V_{be} - j\omega C_\mu V_{cb} \tag{2.181}$$

where I_c, I_b, V_{be}, V_{ce}, and V_{cb} represent the small-signal components of the port variables. The small-signal parameters are defined as follows:

$$r_e \equiv \frac{kT}{q} \frac{1}{I_C} \tag{2.182}$$

$$r_o \cong \frac{V_{AF}}{I_C} \tag{2.183}$$

[6]Note that the actual form of the second term on the right-hand side of (2.180) is V_{cb}/r_o. Since $V_{cb} = V_{ce} - V_{be}$, one can replace this term with V_{ce}/r_o after modifying the first term as $(r_e^{-1} - r_o^{-1})V_{be}$. But, since $r_o \gg r_e$, the first term is practically unchanged.

$$C_\pi \equiv C_{je} + C_{de} \tag{2.184}$$

$$C_\mu \equiv C_{jc} + C_{dc} \tag{2.185}$$

The small-signal ac equivalent circuit representing (2.180) and (2.181) is shown in Figure 2.26(a) together with the parasitic components R_B, R_C, R_E, and C_{cs}.

We can now examine the high-frequency performance of a BJT with the aid of this equivalent circuit. Conventionally, the high-frequency testing is done under the condition of a signal short between the collector and emitter terminals. After ignoring R_E, assuming $R_C \ll r_o$, and calculating the current gain I_c/I_b from the circuit, we arrive at the transfer function

$$\beta(\omega) \equiv \frac{I_c}{I_b} = \frac{\beta(0)}{1 + j\omega\beta(0)r_e(C_\pi + C_\mu)} \tag{2.186}$$

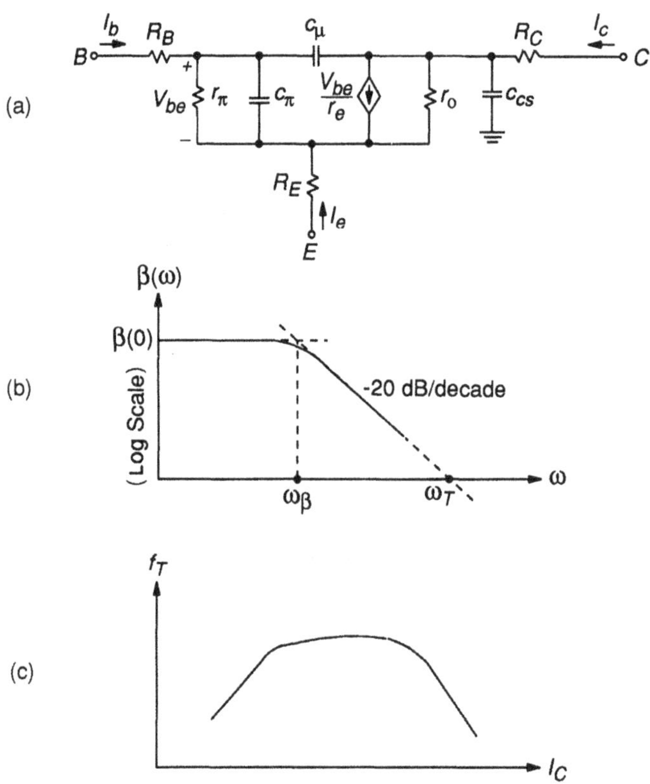

Figure 2.26 (a) The small-signal equivalent circuit representing (2.180) and (2.181). (b) The frequency-dependent variation of β_F. (c) The bias-dependent variation of the transition frequency.

where $\beta(0)$ is the forward beta for $\omega = 0$. As shown in Figure 2.26(b), β rolls off at a rate of 20 dB/decade for those frequencies greater than the so-called *beta cutoff frequency*, which is defined by

$$f_\beta \equiv \frac{1}{2\pi\beta(0)r_e(C_\pi + C_\mu)} = \frac{qI_C}{2\pi kT\beta(0)(C_\pi + C_\mu)} \tag{2.187}$$

The frequency f_T at which the high-frequency asymptote extrapolates to $\beta = 1$ is called the *transition frequency*, which is by far the most popular of the parameters used for characterizing the high-frequency performance of a BJT. It can be expressed with the aid of (2.186) as

$$f_T \equiv \frac{1}{2\pi r_e(C_\pi + C_\mu)} = \frac{qI_C}{2\pi kT(C_\pi + C_\mu)} \tag{2.188}$$

Since

$$f_T = \beta(0)f_\beta \tag{2.189}$$

the bandwidth of the forward beta is precisely equal to the transition frequency.

Taking into the consideration the fact that, in the forward active mode,

$$C_\mu \cong C_{jc}$$

and

$$C_\pi = C_{je} + \tau_F \frac{q}{kT} I_S \exp\left(\frac{q}{kT}V_{BE}\right) = C_{je} + \tau_F \frac{q}{kT} I_C$$

we can rewrite (2.188) as

$$f_T = \frac{1}{2\pi\left[\tau_F + \dfrac{kT}{q}\dfrac{1}{I_C}(C_{je} + C_{jc})\right]} \tag{2.190}$$

As shown in Figure 2.26(c), f_T is proportional to I_C for small values of the latter because τ_F is negligible in comparison with the second term in the denominator of (2.190). At medium and high levels of current, on the other hand, τ_F becomes dominant. At high levels of bias, however, τ_F itself becomes an increasing function of I_C because the Kirk effect or quasisaturation extends the basewidth, which, in turn, increases the base transit time τ_b [see(2.166) and (2.157)]. For this reason, f_T rolls off at high levels of current.

PROBLEMS

2.1 In Section 2.2.5 we presented a detailed physical interpretation for the forward active mode of BJT operation. Give a physical interpretation in similar detail, and construct the counterpart of Figure 2.5 for each of the following modes:
(a) Reverse-active mode: $V_{CB} \ll -kT/q$, $V_{EB} \gg kT/q$.
(b) Cut off mode: $V_{CB} \gg kT/q$, $V_{EB} \gg kT/q$.
(c) Forward saturation mode: $V_{EB} \ll V_{CB} \ll -kT/q$.
(d) Reverse saturation mode: $V_{CB} \ll V_{EB} \ll -kT/q$.

2.2 Derive an expression for each of $n_E(x_E)$, $n_C(x_C)$, $J_{nE}(x_E)$, $J_{nC}(x_C)$, $\phi_{FnB}(x_B)$, $\phi_{FpE}(x_E)$, and $\phi_{FpC}(x_C)$ as a function of bias voltages, terminal currents, structural parameters, physical constants, and, of course, position. Plot a typical profile for each of these internal variables for the forward active mode of operation,

2.3 We introduced the *surface recombination velocity* parameter s in Problem 1.7. Assuming a finite s at the emitter contact, the emitter hole current density can be described by an equation similar to that of part(c) of Problem 1.7. Discuss the effect of a finite s on the Ebers-Moll parameters β_F, β_R, and I_S.

2.4 In deriving the Ebers-Moll model, we assumed $W \ll L_{nB}$, which enabled us to use (1.94) for the electron excess concentration profile $n'_B(x_B)$. If base width W is comparable to L_{nB}, neither (1.94) nor (1.93) is applicable to the analysis of the base region; one has to obtain the general solution of (1.91), which is

$$n'_B = n'_B(0)\cosh(x_B/L_{nB}) + \frac{n'_B(W) - n'_B(0)\cosh(W/L_{nB})}{\sinh(W/L_{nB})}\sinh(x_B/L_{nB})$$

Derive expressions for $J_{nB}(0)$ and $J_{nB}(W)$ from this profile, and reconstruct the Ebers-Moll equations. Discuss the effect on β_F in the forward active mode of operation.

2.5 In a *double-diffused* BJT, the CB junction is formed by implanting and diffusing acceptor-type base dopants into a uniformly n-type doped silicon wafer. A subsequent donor-type emitter implant creates the EB junction as shown in Figure P2.5. These two types of dopants distribute according to a Gaussian concentration profile; that is,

$$N_{DE}(x) = N_{SE}\exp\left[-\left(\frac{x}{\lambda_E}\right)^2\right] \text{ (donor profile)}$$

$$N_{AB}(x) = N_{SB}\exp\left[-\left(\frac{x}{\lambda_B}\right)^2\right] \text{ (acceptor profile)}$$

where N_{SE} and N_{SB} are the surface concentrations, and λ_E and λ_B are two constants determined by processing conditions. The net doping concentration profile is given by

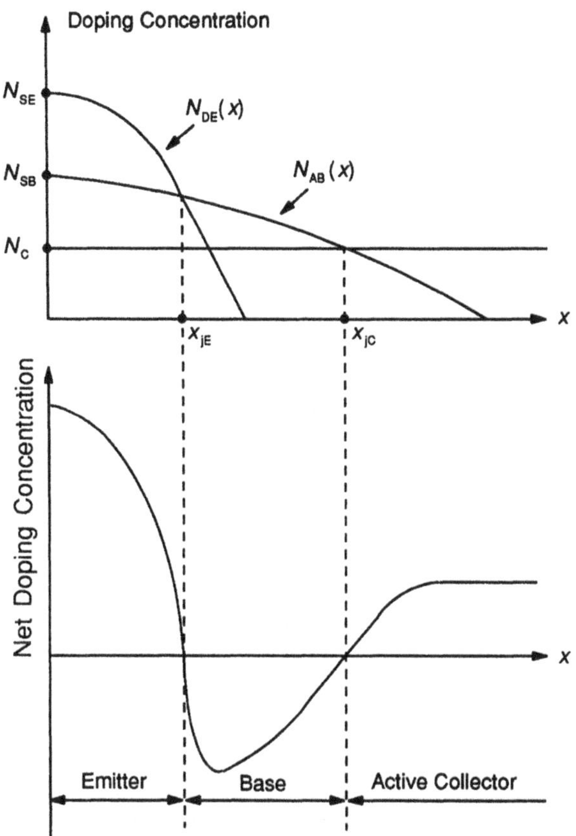

Figure P2.5

$$N(x) = N_{DE}(x) + N_C - N_{AB}(x)$$

where N_C is uniform background donor concentration of the wafer.

(a) For $N_{SE} = 8 \times 10^{20}$ cm^{-3}, $\lambda_E = 0.3$ μm, $N_{SB} = 6 \times 10^{18}$ cm^{-3}, $\lambda_B = 0.55$ μm, and $N_C = 5 \times 10^{15}$ cm^{-3}, calculate the EB junction depth x_{jE} and the CB junction depth x_{jC}. You can either numerically (or graphically) solve the equation $N_{DE}(x) + N_C - N_{AB}(x) = 0$, or obtain an analytical expression for the junction depth by solving $N_{DE}(x) - N_{AB}(x) \cong 0$ for the EB junction where N_C is negligible, and $N_C - N_{AB}(x) \cong 0$ for the CB junction where $N_{DE}(x)$ is negligible. (*Answer:* $x_{jE} = 0.792$ μm, $x_{jC} = 1.47$ μm.)

(b) Numerically (or graphically) calculate the base Gummel number G_b and the emitter effective Gummel number G_e assuming that band gap narrowing in the emitter is described by

$$\Delta E_g = 9\left\{\ln\left(\frac{N}{10^{17}}\right) + \sqrt{\left[\ln\left(\frac{N}{10^{17}}\right)\right]^2 + 0.5}\right\}$$

where N and ΔE_g have units of cm^{-3} and meV, respectively. Ignore transition regions and assume room temperature, for which $kT = 25.8$ meV. (*Answer:* $G_b = 7.85 \times 10^{12}$ cm^{-2}, $G_e = 6.80 \times 10^{13}$ cm^{-2}.)

(c) Assuming an emitter area of 20 × 20 μm, and diffusivities $D_{nB} = 18$ cm^2/s, $D_{pE} = 1.5$ cm^2/s, calculate I_S and β_F. (*Answer:* $I_S = 1.76 \times 10^{-16}$ A, $\beta_F = 104$.)

(d) Calculate the Early voltage V_{AF} for $V_{CB} = 5$V, $V_{CB} = 10$V, and $V_{CB} = 15$V. (*Answer:* $V_{AF} = 123$, 165, and 194V.)

(e) Assuming $D_{pB} = 5$ cm^2/s, calculate the zero-bias base spreading resistance for two-sided base contact. (*Answer:* $R_B = 342\ \Omega$.)

(f) The emitter-base junction capacitance is defined by (2.143). To model C_{je}, one has to find a relationship between G_{be} and v_{EB} in terms of structural parameters. Selecting $x = x_{jE}$ as the origin of a new position variable z, we can express the donor and acceptor profiles around $z = 0$, as

$$N_{DE} = N_{SE} \exp\left[-\left(\frac{z + x_{jE}}{\lambda_E}\right)^2\right]$$

$$N_{AB} = N_{SB} \exp\left[-\left(\frac{z + x_{jE}}{\lambda_B}\right)^2\right]$$

Ignoring the small N_C, substituting these two equations into the net doping concentration equation $N = N_{DE} - N_{AB}$, expanding the latter into Taylor's series around $z = 0$, and ignoring the terms of order 2 and higher, you will obtain

$$N = -az \tag{P2.5.1}$$

where

$$a \equiv 2x_{jE}N_o\left(\frac{1}{\lambda_E^2} - \frac{1}{\lambda_B^2}\right)$$

is the linear grading coefficient and

$$N_o \equiv N_{SE} \exp\left[-\left(\frac{x_{jE}}{\lambda_E}\right)^2\right] = N_{SB} \exp\left[-\left(\frac{x_{jE}}{\lambda_B}\right)^2\right]$$

is the equal acceptor and donor concentrations at the junction. Now G_{be} can be described as

$$G_{be} = \int_0^{z_d} |N|\, dz$$

where z_d is the depletion width on the base side of the junction. Since (P2.5.1) indicates a linear doping concentration profile, the depletion width on the emitter side is also equal to z_d. Assuming $\mathscr{E}(-z_d) = \mathscr{E}(z_d) = 0$, and $\phi(z_d) - \phi(-z_d) = -v_{EB} - V_b$, where V_b is the built-in voltage, show that

$$z_d = \left[\frac{3\epsilon}{2qa}(V_b + v_{EB})\right]^{1/3}$$

$$G_{be} = \frac{a}{2}\left[\frac{3\epsilon}{2qa}(V_b + v_{EB})\right]^{2/3}$$

and

$$C_{je} = \frac{C_{jE}(0)}{[1 + (v_{EB}/V_b)]^{1/3}}$$

where

$$C_{jE}(0) \equiv \frac{A\epsilon}{2[(3\epsilon/2qa)V_b]^{1/3}}$$

Calculate C_{je} for $V_b = 0.70$V and $V_{EB} = -0.75$V. (*Answer:* $C_{je} = 1.2$ pF.)

(g) Calculate the collector-base junction capacitance C_{jC} from (2.150) and (2.84) for $V_b = 0.75$V and $V_{CB} = 5$, 10, and 15V. (*Answer:* $C_{jC} = 34$, 24.9, and 20.6 fF.)

(h) According to the model presented in Section 2.4.1 the forward transit time τ_F can be expressed as

$$\tau_F = \left[1 + \frac{l_E}{W}\frac{N_B(0)}{N_E(0)}\right]\frac{W^2}{2D_{nB}}$$

where $N_B(0)$ and $N_E(0)$ denote the doping concentration at the base and emitter boundaries of the EB transition region. Verify this equation; then, considering the approximate equality of $N_E(0)$ and $N_B(0)$ due to the linear concentration profile around the EB junction, calculate τ_F. (*Answer:* τ_F = 277 ps.)

(i) Calculate the transition frequency f_T from (2.190) for V_{BE} = 0.7V and V_{CB} = 5V. (*Answer:* f_T = 277 MHz.)

2.6 In this exercise you are expected to analyze the Kirk effect on the basis of a simple model. Shown in Figure P2.6(a) is the reference convention for the position variable x and collector current density J_C. Assuming a very high doping concentration in the base and buried-layer region, the electric-field

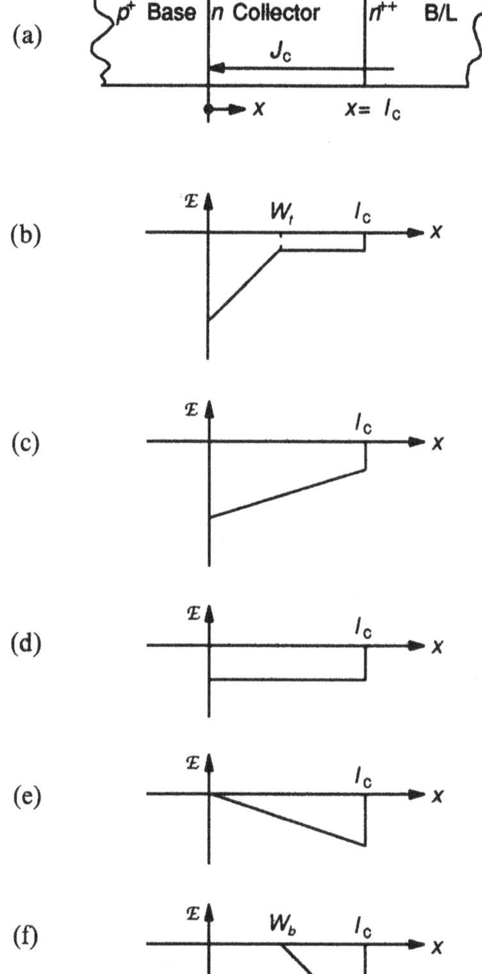

Figure P2.6

profile for relatively low levels of J_C is depicted in Figure P2.6(b). In the space charge region extending between $x = 0$ and $x = W_t$, the dominant charged entities are the donor dopants of concentration N_C and the electrons injected from the base. Assuming a sufficiently strong field in this region, the electrons move by drift at saturation velocity v_s, hence $J_C = qv_s n$. Assuming a sufficiently weak field $\mathscr{E}(W_t)$ in the neutral active collector region, which extends between $x = W_t$ and $x = l_C$, the current density in this region can be expressed as $J_C = q\mu_n N_C \mathscr{E}(W_t)$.

(a) Show that the width of the space-charge region is given by

$$W_t = \sqrt{\frac{2\epsilon_s}{q\left(N_C - \dfrac{J_C}{qv_s}\right)}\left(V_{CB} + V_b - \frac{l_C}{q\mu_n N_C}J_C\right)}$$

where V_b is the built-in voltage for $V_{CB} = 0$ and $J_C = 0$.

(b) The term W_t is an increasing function of V_{CB}. For a specific value of V_{CB} the space charge region occupies the entire collector. This specific value of V_{CB} for $J_C = 0$ is called punch-through voltage V_{pt}. Show that

$$V_{pt} = \frac{qN_C}{2\epsilon}l_C^2 - V_b$$

(c) For $V_{CB} < V_{pt}$, as we increase J_C, the space-charge region either contracts and leads eventually to quasisaturation, or expands and leads to the Kirk effect. Show that the contraction and expansion conditions are given by $V_{CB} < V_K$ and $V_{CB} > V_K$, respectively, where

$$V_K \equiv \frac{l_C}{\mu_n}v_s - V_b$$

(d) Assuming that V_{CB} is fixed at some value in the range $V_K < V_{CB} < V_{pt}$, an increasing J_C will eventually expand the space-charge region all the way to the buried-layer, that is $W_t = l_C$, as shown in Figure P2.6(c). Show that this occurs for

$$J_C = J_{C1} \equiv \frac{V_{pt} - V_{CB}}{l_C\left(\dfrac{l_C}{2\epsilon v_s} - \dfrac{1}{q\mu_n N_C}\right)}$$

(e) Increasing J_C above J_{C1} will eventually lead to a flat-field profile as shown in Figure P2.6(d). Show that this condition occurs when $n = N_C$, and therefore

$$J_C = J_{C2} \equiv qv_sN_C$$

(f) Increasing J_C above J_{C2} causes n to exceed N_C, which reverses the polarity of the slope of the field profile. For $J_C = J_{C3}$, the field at $x = 0$ reduces to zero as shown in Figure P2.6(e). Show that

$$J_{C3} = \frac{2\epsilon v_s}{l_C^2}(V_{CB} + V_b) + qv_sN_C$$

(g) Increasing J_C beyond J_{C3} will increase n, and therefore the negative slope of the field profile, whereas the area bounded by the field profile remains constant because $V_{CB} + V_b$ is constant. Inevitably, therefore, the left-hand boundary of the space-charge region moves into the collector as shown in Figure P2.6(f). Show that

$$W_b = l_C - \sqrt{2\epsilon v_s \frac{V_{CB} + V_b}{J_C - qv_sN_C}}$$

The passage of the left-hand boundary of the space-charge region indicates an extension of the neutral base region, whose total width is now equal to $W + W_b$. This is what is known as the Kirk effect.

(h) An experimental BJT reported by Liu et al. [22] is specified by $N_B = 8 \times 10^{17}$ cm^{-3}, $N_C = 2 \times 10^{16}$ cm^{-3}, $W = 0.04$ μm, and $l_C = 0.6$ μm. Assuming $V_b = 0.76$V, $v_s = 10^7$ cm/s, and finding the appropriate value of μ_n from Figure 1.6, calculate V_{pt} and V_K. (*Answer:* $V_{pt} = 5.54$V, $V_K = -0.245$V.) The authors claim that this device cannot exhibit quasisaturation. Can you justify this claim? For $V_{CB} = 3$V calculate J_{C1}, J_{C2}, and J_{C3}. (*Answer:* $J_{C1} = 1.62 \times 10^4$ A/cm^2, $J_{C2} = 3.2 \times 10^4$ A/cm^2, and $J_{C3} = 5.37 \times 10^4$ A/cm^2.)

(i) The variation of J_C with V_{BE} under the influence of Webster and Kirk effects can be modeled with (2.103) in which W is to be replaced by $W + W_b$, $n_B(0)$ is described by (2.107), and $n_B(W)$ is given by J_C/qv_s. Supposing that the BJT described in part (h) operates at $T = 35$°C, solve (2.103) numerically to obtain the J_C Gummel plot and n_C versus V_{BE} and W_b versus V_{BE} plots. Compare your results with those of Figure 9 in Liu et al. [22]

2.7 The model developed in Section 2.3.1 for the base resistance and emitter current crowding effects is valid only for low-level injection in the base because (2.87) is based on

$$\int_0^W p_B \, dx_B = G_b$$

whose underlying assumption $p_B \equiv N_B$ collapses under high-level injection conditions. Removing the assumption of low-level injection will enable us to model base resistance and current crowding in the presence of the Webster effect and, for that matter, of collector-related high-bias effects as well. Show that, for a uniformly doped base, this leads to the following general form for (2.87):

$$I = -q\mu_{pB}LW\frac{[n_B(0) - n_B(W)][n_B(0) + n_B(W) + N_B]}{2[n_B(0) - n_B(W)] - N_B \ln\{[n_B(0) + N_B]/[n_B(W) + N_B]\}}\frac{dV}{dy}$$

where

$$n_B(0) = -\frac{N_B}{2} + \sqrt{\left(\frac{N_B}{2}\right)^2 + n_i^2 \exp\left(-\frac{q}{kT}V_E\right)\exp\left(\frac{q}{kT}V\right)}$$

Hint: Use the differential equation that follows (2.102) and invoke the neutrality of base.

2.8 Irradiating a BJT with an energetic electron beam results in permanent degradation of β_F. Two physical mechanisms may be responsible for this degradation: (1) *bulk displacement damage*, which effectively increases the basewidth W, or (2) *trap enhancement*, which, by effectively lowering the carrier lifetime in and around the EB transition region, keeps the BJT under a low-current beta roll-off effect all the way up to the onset of high-bias effects [23].

(a) Sketch three β_F versus I_C plots, first for a typical virgin (nonirradiated) BJT, second for a BJT with bulk displacement damage, and third for a BJT with trap enhancement. Discuss the differences and underlying reasons.

(b) The transition frequency f_T is expected to decrease in the case of bulk displacement damage. Why?

(c) Figure P2.8(a) shows the frequency response of β_F before and after irradiation. Can you determine the dominant degradation mechanism?

(d) Which dynamic device parameters can you extract from the experimental data given in Figure P2.8(b)? Assuming room temperature (T = 300K) operation, calculate the value of all you can extract.

2.9 Sakui et al. [24] have proposed a bipolar SRAM cell based on the negative-resistance effect associated with the $I_B - V_{BE}$ characteristics of a BJT. Figure P2.9 shows the I_B Gummel plot obtained for V_{CE} = 1V and V_{CE} = 6V. Apparently, the effect is activated only at relatively large values of V_{CE}. The authors attribute the effect to avalanche multiplication in the CB transition region. If you inspect (2.131) after ignoring the small last term on the right-hand side, you indeed will see the possibility of a negative I_B when an avalanche

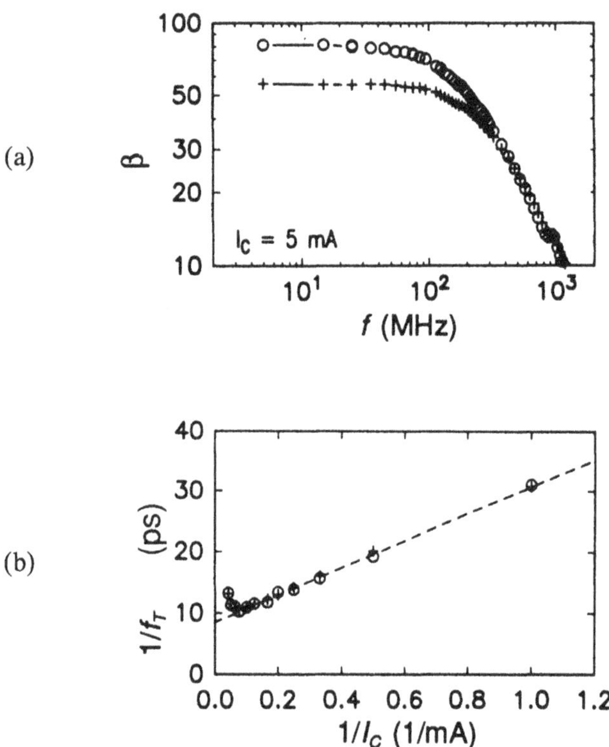

Figure P2.8 (From Jenkins [23]. © 1989 by IEEE. Reprinted with permission of the publisher.)

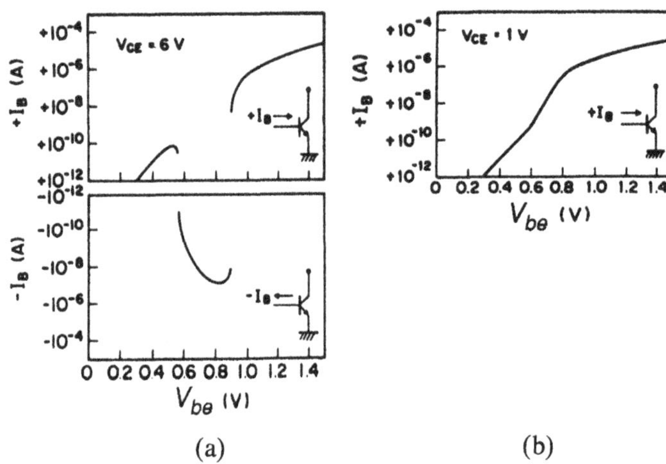

Figure P2.9 (From Sakui [24]. © 1989 by IEEE. Reprinted with permission of the publisher.)

multiplication effect is present. Paying due attention to (1) the low-current β roll-off effect, (2) the variation of peak electric field in the CB transition region with I_C as the device approaches the domain of the Kirk effect (Figure P2.6), and (3) the fact that M is an increasing function of the peak field in the CB transition region, explain why a negative I_B section is present in between two positive I_B sections on the Gummel plot.

2.10 In this exercise we reconsider the charge-control model equations for the purpose of improving the accuracy of the transition frequency expression (2.190). As the first step of improvement, take into account the time variation of the electron charge inside the CB transition region. This charge, which was ignored in the original analysis, can be expressed as

$$Q_{ct} = Aq\int_{x2}^{x_2'} n\,dx$$

Also ignored in the original analysis is the electron charge in the EB transition region. This charge is indeed small. So, we still ignore it.

(a) Show that Q_{ct} can be included in (2.161) as

$$i_C = I_C + C_{jc}\frac{dv_{CB}}{dt} - \frac{d}{dt}(Q_{br} + Q_c + Q_{ct}) \qquad \text{(P2.10.1)}$$

whereas (2.160) remains the same, that is,

$$i_E = I_E - C_{jc}\frac{dv_{BE}}{dt} - \frac{d}{dt}(Q_{bf} + Q_e) \qquad \text{(P2.10.2)}$$

(b) As the next step of improvement, suppose that the BJT operates in the forward active mode with a sufficiently large reverse bias across the CB transition region, so that electrons drift at the saturation velocity in that region. This enables you to assume $n = i_C/Aqv_s$ in the transition region including the boundaries at $x = x_2$ and $x = x_2'$. Derive the equations

$$Q_{br} = \frac{W}{2v_s}i_C$$

and

$$Q_{ct} = \frac{W_{tc}}{v_s}i_C$$

where W_{tc} is the width of the CB transition region. Based on the assumption of quasistatic operation in the base, use

$$i_C = Aq\frac{D_{nB}}{W}[n'_B(x_1) - n'_B(x_2)]$$

to obtain an expression for $n'_B(x_1)$ in terms of i_C. This will lead you to

$$Q_{bf} = \left(\frac{W^2}{2D_{nB}} + \frac{W}{2v_s}\right)i_C$$

and, assuming $p'_E(x'_1) \cong n'_B(x_1)$ for the narrow and linearly graded EB transition region, also to

$$Q_e = \frac{l_e}{W}Q_{bf}$$

Also persuade yourself that $Q_c \cong 0$ in the forward active mode.

(c) Now derive from (P2.10.1) and (P2.10.2) the following small-signal characteristic equations:

$$I_c = \frac{1}{r_e}V_{be} + \frac{1}{r_o}V_{cb} + j\omega C_{jc}V_{cb} - j\omega(C_{br} + C_{ct})V_{be}$$

$$I_b = \frac{1}{\beta_F r_e}V_{be} - j\omega C_{jc}V_{cb} + j\omega(C_{br} + C_{ct} + C_{je} + C_{bf} + C_e)V_{be}$$

where the capacitances, except for C_{jc} and C_{je}, are defined as

$$C \equiv \frac{\partial Q}{\partial i_C}\frac{di_C}{dv_{BE}}$$

and are identified with the same subscripts as the charges with which they are associated.

(d) Write an expression for short-circuit current gain $\beta(\omega) = I_c/I_b|_{v_{ce}=0}$, and determine f_T as in (2.190). The forward transit time τ_F in your improved expression will now be given by

$$\tau_F = t_{bb} + t_{ct} + t_e$$

where

$$t_{bb} \equiv \frac{W^2}{2D_{nB}} + \frac{W}{v_s}$$

is the total base delay,

$$t_{ct} \equiv \frac{W_{tc}}{v_s}$$

is the CB transition region delay, and

$$t_e = \frac{l_E W}{2D_{nB}} + \frac{l_E}{2v_s}$$

is the emitter delay.

2.11 We introduced the concept of band gap narrowing in Section 2.2.6 in connection with the emitter Gummel number. In practice band gap narrowing may occur in the base region as well. Generally, therefore, β_F in a uniformly doped BJT can be expressed as

$$\beta_F = \frac{N_E l_E \exp(-\Delta E_{gE}/kT)}{N_B W \exp(-\Delta E_{gB}/kT)}$$

where N_E and N_B are the actual doping concentrations in the emitter and base, respectively, and ΔE_{gE} and ΔE_{gB} denote band gap narrowing in these two regions. Since ΔE_{gE} and ΔE_{gB} are increasing functions of the doping concentration, and the doping concentration is higher in the emitter, we expect $\Delta E_{gE} > \Delta E_{gB}$.

(a) Assuming $\Delta E_{gE} - \Delta E_{gB}$ to be independent of the temperature, find an expression for the temperature sensitivity of β_F, that is, $d\beta_F/dT$. Then show what happens to β_F as the operation temperature increases. Is your conclusion supported by the experimental data of Figure P2.11?

(b) The data presented in Figure P2.11 were published by Cressler et al. [25]. The room temperature behavior is typical in the sense that β_F remains relatively constant for moderate levels of current and then rolls off due to a Kirk effect at high currents. What is unusual is the low-temperature behavior: β_F increases with I_C and reaches a peak before the Kirk effect becomes dominant. The authors explain this phenomenon by postulating an *injection-induced bandgap narrowing* in the base, which makes ΔE_{gB} an increasing function of I_C. Based on this information, write an expression for the "relative current-sensitivity of β_F" as defined by $\partial\beta_F/\beta_F\partial I_C$. It must be expressed in terms of $\partial\Delta E_{gB}/\partial I_C$, which is the rate of injection-induced bandgap narrowing, and of $\partial W/W\partial I_C$, which is the relative rate of Kirk effect. Now explain why a peak occurs in β_F versus I_C curves only at low temperatures.

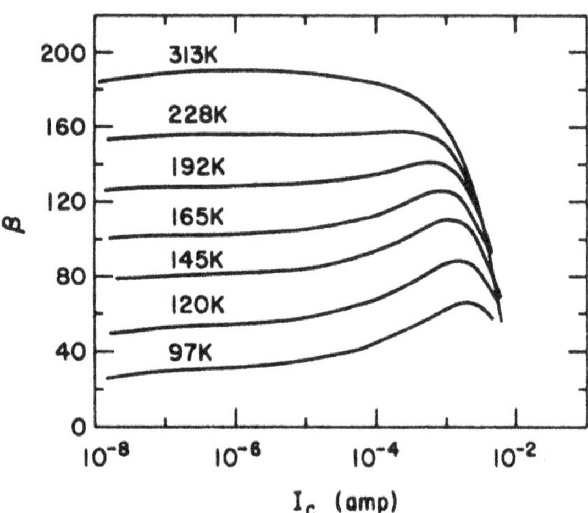

Figure P2.11 (From Cressler [25]. © 1989 by IEEE. Reprinted with permission of the publisher.)

2.12 It is possible to improve the speed and current gain of a BJT by fabricating the emitter with polysilicon material. As shown in Figure P2.12, the base and collector regions of a polyemitter BJT are built in monocrystalline silicon as in conventional devices. The emitter comprises a very narrow monocrystalline

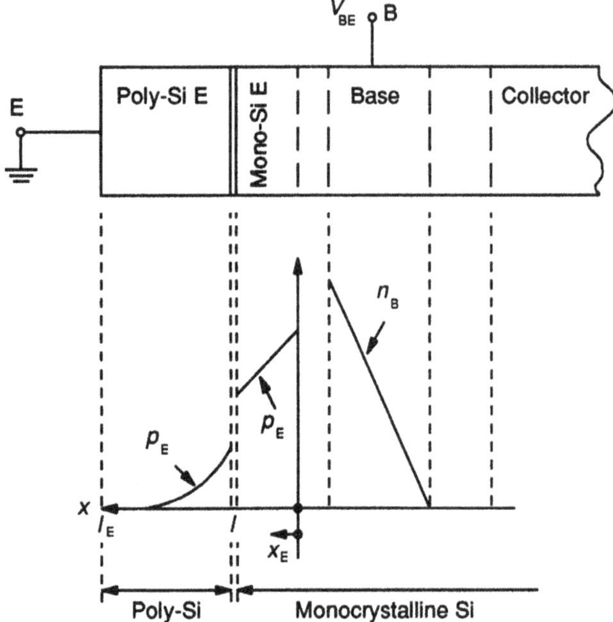

Figure P2.12

segment and a polycrystalline bulk, which are separated by a thin interfacial oxide layer located at $X_E = l$.

(a) Following the methodology of Section 2.2.6, show that the hole current density at $x_E = 0$ can be described by

$$J_{pE}(0) = \frac{qD_{pE}n_i^2}{G_e + (n_i/n_{ie})^2[(D_{pE}L_{pp})/(sL_{pp} + \alpha D_{pp})]N_E(l)}\left[\exp\left(\frac{q}{kT}V_{BE}\right) - 1\right]$$

where D_{pE}, n_{ie}, and G_e are, respectively, the hole diffusivity, intrinsic carrier concentration, and effective Gummel number in the monocrystalline emitter region; $N_E(l)$ is the doping concentration at $x_E = l$, where the interface is located; L_{pp} and D_{pp} are, respectively, the diffusion length and diffusivity of holes in the polycrystalline emitter region, which extends between $x = l$ and $x = l_E$; s is the surface recombination velocity at the interface at $x_E = l$ [you need to relate $J_p(l^+)$ to $J_p(l)$ through s and $p_E'(l)$]; and α is an interfacial factor defined as $\alpha \equiv p_E'(l^+)/ p_E'(l)$. In your derivation, assume l to be much shorter than the minority-carrier diffusion length in the monocrystalline emitter, and $l_E - l$ to be much longer than L_{pp}.

(b) Compare the equation of $J_{pE}(0)$ with equation (16) of Suzuki [26]. Determine the differences and discuss the reasons behind these differences.

REFERENCES

[1] Ghandi, Sorab K., *VLSI Fabrication Principles*, New York: John Wiley & Sons, 1983, Sec. 11.9.

[2] Roulston, David J., *Bipolar Semiconductor Devices*, New York: McGraw-Hill, 1990, Chapters 7-9.

[3] Warner, Jr., R. M., and B. L. Grung, *Semiconductor-Device Electronics*, Philadelphia: Holt, Rinehart and Winston, 1991, Chap. 4.

[4] Antognetti, P., and G. Massobrio, *Semiconductor Device Modeling with SPICE*, New York: McGraw-Hill, 1988, Chap. 2.

[5] Ebers, J. J., and J. L. Moll, "Large-Signal Behavior of Junction Transistors," *IRE Proc.*, Vol. 42, 1954, p. 1761.

[6] Gummel, H. K., "Measurement of the Number of Impurities in the Base Layer of the Transistor," *IRE Proc.*, Vol. 49, 1961, p. 834.

[7] Early, J. M., "Effects of Space-Charge Layer Widening in Junction Transistors," *IRE Proc.*, Vol. 40, 1952, p. 1401.

[8] Hauser, J. R., "The Effects of Distributed Base Potential on Emitter-Current Injection Density and Effective Base Resistance for Stripe Transistor Geometries," *IEEE Trans. on Electron Devices*, Vol. ED-11, 1964, p. 238.

[9] Lary, Janifer E., and R. L. Anderson, "Effective Base Resistance of Bipolar Transistors," *IEEE Trans. on Electron Devices,* Vol. ED-32, 1985, p. 2503.

[10] Myungsuk, Jo, and Dorothea E. Burk, "An Intrinsic Base Resistance Model for Low and High Currents," *IEEE Trans. on Electron Devices*, Vol. 37, 1990, p. 202.

[11] Webster, W. M., "On the Variation of Junction-Transistor Current-Amplification Factor with Emitter Current," *IRE Proc.,* Vol. 42, 1954, p. 914.

[12] Bowler, David L., and Fredrik A. Lindholm, "High Current Regimes in Transistor Collector Regions," *IEEE Trans. on Electron Devices*, Vol. ED-20, 1973, p. 257.

[13] Rey, G., F. Dupuy, and J. P. Bailbe, "A Unified Approach to the Base Widening Mechanisms in Bipolar Transistors," *Solid-State Electron.*, Vol. 18, 1975, p. 863.
[14] Grung, B. L., and R. M. Warner, Jr., "An Analytical Model for the Epitaxial Bipolar Transistor," *Solid-State Electron.*, Vol. 20, 1977, p. 753.
[15] Hanggeun, J., and Jerry G. Fossum, "A Charge-Based Large-Signal Bipolar Transistor Model for Device and Circuit Simulation," *IEEE Trans. on Electron Devices*, Vol. 36 1989, p. 124.
[16] Kirk, Jr., C. T., "A Theory of Transistor Cutoff Frequency (f_t) Falloff at High Current Densities," *IRE Trans. on Electron Devices,* Vol. 9, 1962, p. 164.
[17] Moll, J. L., *Physics of Semiconductors*, New York: McGraw-Hill, 1964, Chap. 11 and 12.
[18] Liou, J. J., and J. S. Yuan, "An Avalanche Multiplication Model for Bipolar Transistors," *Solid-State Electron,* Vol. 33, 1990, p. 35.
[19] Roulston, D. J., and M. Depey, "Emitter-Collector Breakdown Voltage BV_{CEO} versus Gain h_{fe} for Various NPN Collector Doping Levels," *Electron. Lett.*, Vol. 16 1980, p. 803.
[20] Warner, Jr., R. M., and B. L. Grung, *Semiconductor-Device Electronics*, Philadelphia: Holt, Rinehart and Winston, 1991, Sec. 4.7 and 4.8.
[21] Antognetti, P., and G. Massobrio, *Semiconductor Device Modeling with SPICE*, New York: McGraw-Hill, 1988, Sec. 2.5 and 2.6.
[22] Liu, Mark, Tzu-Yin Chiu, Vance D. Archer III, and Helen H. Kim, "Characteristics of Impact Ionization Current in the Advanced Self-Aligned Polysilicon-Emitter Bipolar Transistor," *IEEE Trans. on Electron Devices*, Vol. 38, 1991, p. 1845.
[23] Jenkins, K. A., "Frequency Response of Bipolar Junction Transistors after Electron-Beam Irradiation," *IEEE Trans. on Electron Devices*, Vol. 36, 1989, p. 1722.
[24] Sakui, K., T. Hasegawa, T. Fuse, S. Watanabe, K. Ohuchi, and F. Masuoka, "A New Static Memory Cell Based on the Reverse Base Current Effect of Bipolar Transistors," *IEEE Trans. on Electron Devices*, Vol. 36, 1989, p. 1215.
[25] Cressler, J. D., D. D.-L. Tang, and E. S. Yang, "Injection-Induced Bandgap Narrowing and its Effects on the Low-Temperature Operation of Silicon Bipolar Transistors," *IEEE Trans. on Electron Devices*, Vol. 36, 1989, p. 2576.
[26] Suzuki, Kunihiro, "Unified Minority-Carrier Transport Equation for Polysilicon or Heteromaterial Emitter Contact Bipolar Transistors," *IEEE Trans. on Electron Devices*, Vol. 38, 1991, p. 1868.

Chapter 3

Metal-Oxide-Semiconductor Transistor

3.1 BASIC MOSFET STRUCTURE

As shown in Figure 3.1, the basic silicon substructure of the metal-oxide semiconductor transistor (MOSFET) consists of two n^+p junctions built on a p-type silicon substrate and a surface space-charge region between the two. A conductive *gate* electrode, made of a suitable metal or highly doped polycrystalline silicon, and an underlying film of insulating SiO_2 form a superstructure by which the transversal boundary field $\mathscr{E}_x(0)$ at the surface can be controlled. The two junctions, being nonforward biased, are not electrically interconnected unless this field induces a

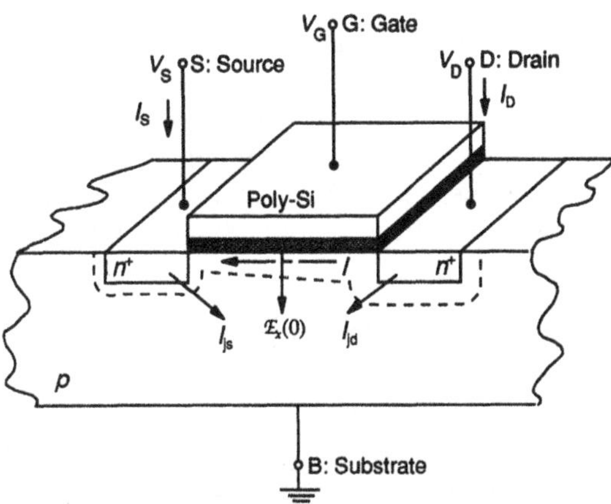

Figure 3.1 A schematic of the n-channel MOSFET structure.

sufficiently large population of electrons in the surface space-charge region. When such a *channel* of electrons is present, an external bias between the two junctions creates a channel current I. This current can be modulated by the gate-substrate voltage because the latter controls $\mathscr{E}_x(0)$, which, in turn, controls the electron concentration in the channel. For structural details and fabrication of MOSFETs, particularly those of the popular CMOS family, the reader is encouraged to consult Chen [1].

The structure depicted in Figure 3.1 is called an *n-channel* MOSFET because the channel charge is due to electrons. Alternatively, a *p-channel* MOSFET is obtained by forming two p^+n junctions on an n-type substrate.

It is obvious from Figure 3.1 that a MOSFET is accessible via four terminals; namely, the *gate* (G), *substrate* (B), *source* (S), and *drain* (D). Selecting the substrate as the reference terminal, we can define three independent port voltages called *source voltage* V_S, *drain voltage* V_D, and *gate voltage* V_G.

Unlike the emitter and collector of a BJT, the source and drain terminals of a MOSFET cannot be distinguished on the basis of structural properties. They are identified solely by the relative magnitudes of their respective bias voltages. In an n-channel MOSFET, the drain is defined as the terminal with more positive voltage. In a p-channel MOSFET, its definition is just the opposite. Therefore, the following conditions are always valid:

$$V_D \geq V_S, \qquad n\text{-channel} \tag{3.1}$$

$$V_D \leq V_S, \qquad p\text{-channel} \tag{3.2}$$

3.2 MOSFET UNDER BIAS

In this section, we analyze an n-channel MOSFET under the most general bias conditions of nonzero V_S, V_D, and V_G. After setting up a foundation for the analysis, we examine the physics of the surface space-charge region in detail and then proceed to device modeling. The model equations of a p-channel MOSFET will be the last topic of discussion in this section. Those who need more depth in these topics will greatly benefit from reading Tsividis [2].

3.2.1 Fundamentals of Nonequilibrium Analysis

A MOSFET with unbiased source and drain junctions ($V_S = 0$, $V_D = 0$) remains in a thermal equilibrium state as long as the gate bias is time invariant. A dc gate bias cannot upset the equilibrium state because the insulating SiO_2 blocks conduction through the gate terminal. On the other hand, a biased source or drain junction conducts current and thus upsets the equilibrium state. Assuming an

n-channel MOSFET in a dc nonequilibrium state, and denoting the source and drain terminal currents by I_S and I_D, respectively, we can express these currents generally as

$$I_D = I + I_{jD} \tag{3.3}$$

$$I_S = -I + I_{jS} \tag{3.4}$$

where I is the *channel current* flowing between the two junctions, and I_{jD} and I_{jS} are the two *substrate currents* associated with the junctions. The substrate currents, which flow outside the channel, are not controllable by the gate and therefore must be minimized. This is why the source and drain junctions are not forward biased under normal operating conditions. For the reverse-biasing positive values of V_S and V_D, these substrate currents result mostly from thermal generation in the space-charge region surrounding the junctions and the surface, and to a lesser extent from the diffusion of electrons from substrate to junctions. Since these conduction processes are of very limited magnitude, the substrate currents are usually negligible in comparison with the channel current. Therefore, we can simplify (3.3) and (3.4) to

$$I_D = -I_S = I$$

This reduces the task of dc steady-state modeling of a MOSFET to that of finding a relationship between the channel current I and the port voltages V_G, V_S, and V_D.

Figure 3.2 shows the reference convention to be used in the analysis. Since the junctions are not forward biased, we do not expect any significant hole current in the lateral and transversal directions, that is

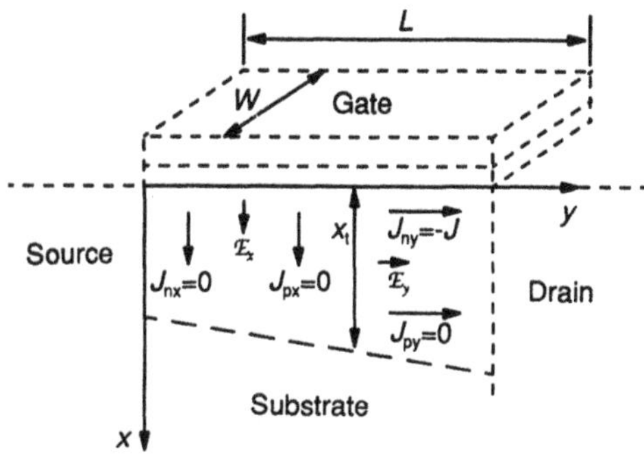

Figure 3.2 The reference convention used in MOSFET analysis.

$$J_{py} = q\mu_p p\mathscr{E}_y - qD_p\frac{\partial p}{\partial y} \cong 0 \tag{3.5}$$

$$J_{px} = q\mu_p p\mathscr{E}_x - qD_p\frac{\partial p}{\partial x} \cong 0 \tag{3.6}$$

The electron current is negligible in the transversal direction as discussed previously in connection with the substrate currents, but is equal to the channel current in the lateral direction. Therefore,

$$J_{ny} = q\mu_n n\mathscr{E}_y + qD_n\frac{\partial n}{\partial y} = -J \tag{3.7}$$

$$J_{nx} = q\mu_n n\mathscr{E}_x + qD_n\frac{\partial n}{\partial x} \cong 0 \tag{3.8}$$

where J is the density of the channel current. Using (3.5), we can write the following expression for the lateral electric field:

$$\mathscr{E}_y \cong \frac{kT}{q}\frac{1}{p}\frac{dp}{dy} \tag{3.9}$$

Substituting this into (3.7), and recognizing that the channel current is described by

$$I = W\int_0^{x_t} J\,dx \tag{3.10}$$

where W and x_t are, respectively, the width and thickness of the surface space-charge region, we arrive at the equation

$$I \cong -W\frac{kT}{q}\int_0^{x_t} q\mu_n n\frac{d\ln(pn)}{dy}dx \tag{3.11}$$

Now, eliminating $\mathscr{E}_x$ between (3.6) and (3.8), we obtain

$$\frac{d\ln(pn)}{dx} \cong 0 \tag{3.12}$$

that is, $\ln(pn)$ is independent of x. Therefore, the derivative $d\ln(pn)/dy$ in (3.11) must also be independent of x. This enables us to rewrite (3.11) as

$$I = -W\frac{kT}{q}\frac{d\ln(pn)}{dy}\int_0^{x_t} q\mu_n n\,dx \tag{3.13}$$

The electron mobility μ_n appearing in (3.13) is generally dependent on x because of the additional scattering mechanisms involved at the interface and also of the x-dependent electron concentration. Nevertheless, defining an *effective electron mobility* as

$$\overline{\mu}_n = \frac{\int_0^{x_t} \mu_n n \, dx}{\int_0^{x_t} n \, dx} \tag{3.14}$$

we can transform (3.13) into

$$I = -W\frac{kT}{q}\overline{\mu}_n\frac{d\ \ln(pn)}{dy}Q_n \tag{3.15}$$

where

$$Q_n \equiv \int_0^{x_t} qn \, dx \tag{3.16}$$

is the magnitude of the electron charge density per unit area in the surface space-charge region. This quantity is called the *channel charge density*.

Now, we once more turn our attention to the product pn. Using (1.22) and (1.23), we can write

$$pn = n_i^2 \exp\left(-\frac{q}{kT}V\right) \tag{3.17}$$

where

$$V \equiv \phi_{Fn} - \phi_{Fp} \tag{3.18}$$

is called the *channel potential*. Using (3.17) in (3.15), we finally arrive at

$$I = W\overline{\mu}_n Q_n \frac{dV}{dy} \tag{3.19}$$

which is the basic differential equation of MOSFET modeling. To develop a model fully, however, we must (1) obtain a relationship between Q_n and V, (2) find appropriate boundary values for V, and (3) integrate (3.19) with the former two. The first two steps of this task call for a more detailed analysis of the surface space-charge region. This is what we will do next.

3.2.2 Analysis of the Surface Space-Charge Region

The composition of the electric charge in the surface space-charge region is determined by the transversal and lateral electric fields, which, in turn, are determined by the three bias voltages V_G, V_S, and V_D together with certain structural parameters. Depending on these bias values, one of the following five conditions prevails in the surface space-charge region:

1. *Accumulation*, which is defined as the case in which the surface concentration of the majority carriers of the substrate (holes in an n-channel MOSFET) is higher than the doping concentration.
2. *Flatband*, which is the case of the majority-carrier surface concentration being equal to the doping concentration.
3. *Depletion*, which represents the case of both electron and hole surface concentrations being lower than the doping concentration.
4. *Threshold*, which is the condition in which minority carriers (electrons in the case of an n-channel MOSFET) have the same concentration as the dopants at the surface.
5. *Inversion*, in which the surface concentration of minority carriers exceeds the doping concentration.

Note that the accumulation, depletion, and inversion conditions involve a range of carrier concentration values, whereas the flatband and threshold conditions are identified with a single value of a certain carrier concentration. Thus, these two conditions demarcate the former three.

Since the channel current of a MOSFET is proportional to the channel charge Q_n of minority carriers [see (3.19)], a large current can be realized only under inversion conditions. The current admitted by the threshold and depletion conditions is relatively small but it may still be significant in some circuit applications. The flatband and accumulation conditions, on the other hand, do not allow for a sufficient channel charge to yield any significant channel current. Although inversion, threshold, and depletion are the only conditions of significance in the current-voltage modeling of a MOSFET, we will now present a general analysis of the surface space-charge region including all five conditions. This will help the reader gain full insight into the physics of semiconductor surfaces.

Transversal Energy-Band Diagram

A transversal energy-band diagram of the MOSFET structure of Figure 3.2 is the back-bone of the analysis we will present for the surface space-charge region. Shown in Figure 3.3(a) is such a diagram drawn for an arbitrary lateral location y for $V_G = V_S = V_D = 0$. Notice the following features.

1. The structure, being unbiased, is in an equilibrium state. Therefore, a single position-independent Fermi level E_F can describe carrier concentrations everywhere in the structure.

2. The polycrystalline (poly) gate and the monocrystalline substrate, both being made of silicon, have the same bandgap potential ϕ_g and the same electron affinity κ_s. The SiO_2 gate insulator has a larger bandgap and a different electron affinity κ_o. These differences cause band discontinuities to appear at $x = -T_{ox}$ and $x = 0$, where the gate-oxide and oxide-silicon interfaces are located. The potential difference corresponding to these discontinuities is denoted by ϕ_{sox} on the energy-band diagram, and is usually referred to as the *Si-SiO₂ work function*. It is obvious from the energy-band diagram that

$$\phi_{sox} = \frac{\kappa_S - \kappa_O}{q}$$

3. The bands are flat in the neutral bulk region of the substrate, but are bent in the surface space-charge region. In terms of an electrostatic potential difference, the bending is denoted by ψ_s, which is referred to as *surface potential*.
4. Generally, an electrostatic potential difference is expected to develop across the oxide, as well. This is denoted by V_{ox}, and is called *oxide voltage*.
5. Although a surface space-charge region is also expected to develop in the gate, the very heavy doping concentration of the poly material does not allow for a significant bending of bands in that space-charge region. This is why the bands are flat in the gate.

According to the potential energy curve $U(x)$ of Figure 3.3(a), the total electrostatic potential difference between the gate and substrate terminals can be described as

$$\psi_G = V_{ox} + \psi_s \tag{3.20}$$

On the other hand, the loop equation

$$\phi_{FG} + \phi_g/2 + \phi_{sox} + V_{ox} - \phi_{sox} - \phi_g/2 + \psi_s - \phi_{FB} = 0$$

written from the same energy-band diagram, yields $V_{ox} + \psi_s = \phi_{FB} - \phi_{FG}$, which enables us to rewrite (3.20) as $\psi_G = \phi_{FB} - \phi_{FG}$. According to this equation, ψ_G in a totally unbiased MOS structure can be regarded as a *built-in voltage*, which is determined solely by the doping concentrations of the gate and substrate regions.

Now suppose $V_S = V_D = 0$ but $V_G \neq 0$. Assuming that the gate insulator does not permit a current flow into or out of the substrate, the latter should still remain under equilibrium conditions and still be characterized by a single position-independent Fermi level, as shown in Figure 3.3(b). However, the externally applied gate voltage is now superimposed on the built-in voltage to alter the electrostatic potential between the gate and substrate terminals as follows:

$$\psi_G = V_G + \phi_{FB} - \phi_{FG} \tag{3.21}$$

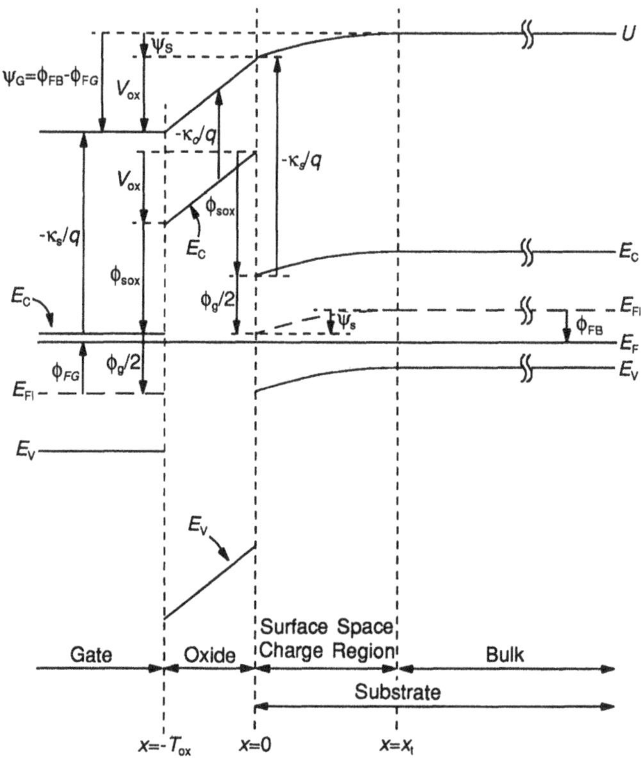

Figure 3.3(a) A transversal energy-band diagram of the MOSFET structure.

As a result, the bands of the gate are displaced vertically by precisely one V_G. This is a downward displacement for $V_G > 0$, and an upward one for $V_G < 0$. Since the position of E_F with respect to the bands in the gate region is fixed by the structural invariants ϕ_{FG} and $\phi_g/2$, E_F must also be displaced by the same amount. As a result, the Fermi levels E_{FG} of the gate and E_{FB} of the substrate are separated by a potential difference of V_G, as depicted in Figure 3.3(b). Also note that, since ψ_G is still the sum of V_{ox} and ψ_s, as described by (3.20), the bias V_G must be absorbed by these two potentials. This fact can be analytically expressed with the aid of (3.20) and (3.21) as follows:

$$V_G = V_{ox} + \psi_s - \phi_{FB} + \phi_{FG} \tag{3.22}$$

As the final step of constructing the energy-band diagram of a MOSFET, we must incorporate the effect of nonzero V_S and V_D, which definitely force the

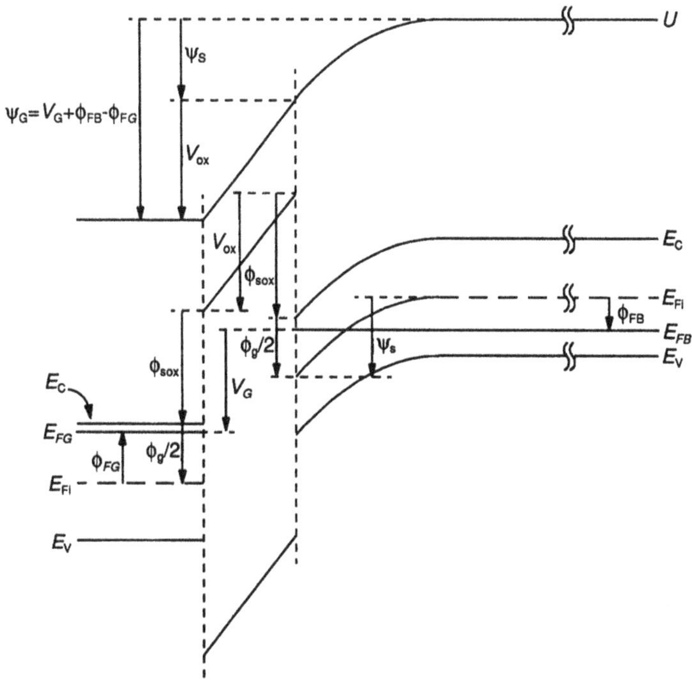

Figure 3.3(b) Splitting in a MOSFET substrate.

substrate into a nonequilibrium state by inducing current. In a nonequilibrium state, generally, the Fermi level must split into two quasi-Fermi levels. Figure 3.3(c) shows the way in which the splitting occurs in a MOSFET substrate. First, note that the ohmic contact at the far end ($x \to \infty$) of the substrate prevents any excess concentration buildup at that boundary, where, in consequence, the two quasi-Fermi levels should merge into a single level. Secondly, the majority-carrier (hole) quasi-Fermi level must be position independent in the quasi neutral bulk, as can be deduced from (1.107) for $J_{px} \cong 0$ and $p \cong N_A$. In the surface space-charge region, p may be small under any of the inversion, threshold, and depletion conditions but, since $J_{px} \cong 0$, it is reasonable to expect the variation of E_{Fp} along the space-charge region to be negligibly small. On the basis of these considerations, E_{Fp} is drawn in Figure 3.3(c) as a straight line coinciding with the substrate Fermi level E_F of the previous two energy-band diagrams. Therefore, the external gate bias V_G now defines the separation between the E_{FG} of the gate and the E_{Fp} of the substrate. For those who wonder why no Fermi level splitting is indicated on the gate side, it suffices to consider the fact that the gate cannot conduct current

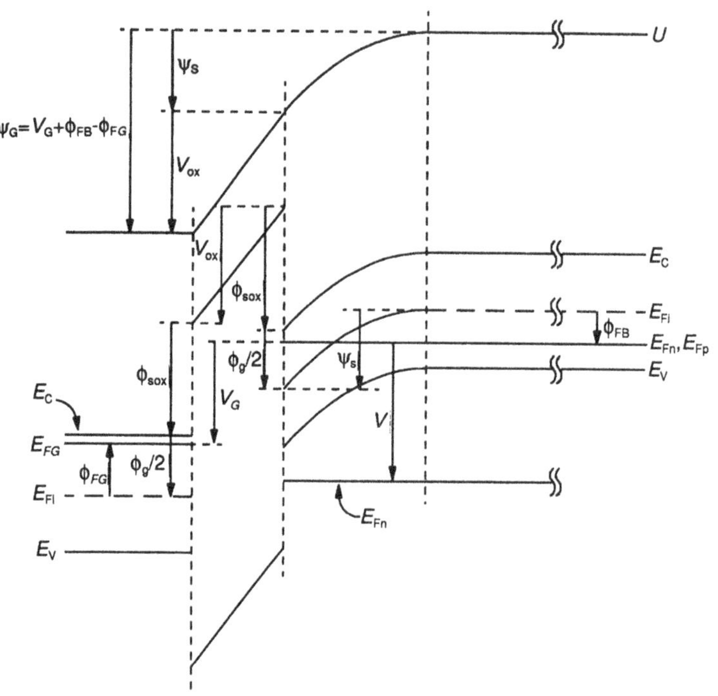

Figure 3.3(c) The MOSFET gate and substrate separated by potential difference.

under steady-state conditions regardless of the bias conditions of the source and drain.

As to the positioning of the minority-carrier (electron) quasi-Fermi level in the substrate, we expect on the basis of (1.108) a very slowly varying E_{Fn} along x because, although the minority-carrier concentration n is unconditionally small in the bulk, and can be small in the space-charge region when a depletion or accumulation condition prevails, the transversal electron current J_{nx} is also very small. For this reason, the variation of E_{Fn} from one boundary of the space-charge region to the other is generally insignificant. This also leads to the conclusion that the channel potential V, which is defined by (3.18) as the potential difference corresponding to the separation between E_{Fn} and E_{Fp} in the space-charge region, is virtually independent of x. As a matter of fact, one can reach this conclusion directly from (3.12), which was written under the assumption of negligible transversal currents.

Among the parameters indicated on the general energy-band diagram of Figure 3.3(c), only V_{ox}, ψ_s, and V can vary with position along the lateral (y) direction because the rest of the parameters, with the exception of V_G, are structural invariants, and V_G is an equipotential for all lateral locations in the gate region.

Actually, lateral variations are of secondary importance in MOSFET current-voltage modeling; the boundary values at $y = 0$ and $y = L$, particularly those of the channel potential V, are of much greater importance. Recalling (1) from (3.18) that V is the potential difference between E_{Fn} and E_{Fp}, (2) from Figure 3.2 that the boundaries $y = 0$ and $y = L$ belong to the transition regions of the source and drain junctions, and (3) from Figure 2.4 that the potential difference between E_{Fn} and E_{Fp} in a *pn* junction transition region equals the junction bias voltage applied to the *n* side with respect to the *p* side, we can reach the conclusion

$$V|_{y=0} = V_S \tag{3.23}$$

$$V|_{y=L} = V_D \tag{3.24}$$

Surface Potential Versus Transversal Boundary Field and Gate Voltage

Figure 3.3(c) makes it obvious that the positioning of the energy bands at $x = 0$ is determined by the surface potential ψ_s. This makes ψ_s a valuable tool for analyzing the surface space-charge region in terms of bias voltages. Note that (3.22) already relates ψ_s to the gate bias V_G. However, this relationship is not explicit in ψ_s because V_{ox} is not independent of the latter. We will further develop (3.22) with a view to obtaining an explicit relationship between ψ_s, V_G, and V. In the process, we will also identify the role of the transversal boundary field $\mathscr{E}(0) \equiv \mathscr{E}_x|_{x=0}$ in the formation of a surface space-charge region.

First, consider the transversal field $\mathscr{E}_{ox}$ inside the gate insulator. If the insulator is assumed to be charge-free, an application of Poisson's equation will yield an x-independent $\mathscr{E}_{ox}$. Writing $d\psi/dx = -\mathscr{E}_{ox}$ and integrating it between $x = -T_{ox}$ and $x = 0$, we obtain the following expression for the oxide potential:

$$V_{ox} = \psi(-T_{ox}) - \psi(0) = T_{ox}\mathscr{E}_{ox} \tag{3.25}$$

where $\mathscr{E}_{ox}$, being the transversal field on the oxide side of the SiO_2-Si interface, can be related to the transversal boundary field $\mathscr{E}(0)$ on the silicon side through a rule of the electromagnetic theory, which states that if an interface of two dielectric media, located for example at $x = 0$, contains a two-dimensional sheet charge of density Q in coulombs per square centimeter, then, the normal component of the displacement vector has a discontinuity of $D(0^+) - D(0^-) = Q$ at that interface. The SiO_2-Si interface of the MOS structures does indeed contain such a sheet charge, which is incorporated inevitably during the processing of SiO_2 in device fabrication. It is called the *oxide fixed charge,* and has a positive density denoted mostly by Q_{ss}. Applying the preceding rule to this interface, and using (1.2) for D, we obtain

$$\mathscr{E}_{ox} = \frac{\epsilon_s}{\epsilon_{ox}}\mathscr{E}(0) - \frac{Q_{ss}}{\epsilon_{ox}} \tag{3.26}$$

where ϵ_s and ϵ_{ox} denote the dielectric constants of the silicon and oxide regions, respectively. Note that ϵ_{ox} is about 3.5×10^{-13} F/cm for SiO_2. Substituting (3.26) into (3.25) and rearranging the latter, we obtain

$$V_{ox} = \frac{\epsilon_s}{C_{ox}}\mathscr{E}(0) - \frac{Q_{ss}}{C_{ox}} \tag{3.27}$$

where

$$C_{ox} \equiv \frac{\epsilon_{ox}}{T_{ox}} \tag{3.28}$$

is the *oxide capacitance per unit gate area* in farads per square centimeter. Finally, using (3.27) in (3.22), we arrive at the equation

$$V_G = V_{FB} + \psi_s + \frac{\epsilon_s}{C_{ox}}\mathscr{E}(0) \tag{3.29}$$

where

$$V_{FB} \equiv \phi_{FG} - \phi_{FB} - \frac{Q_{ss}}{C_{ox}} \tag{3.30}$$

represents the combined effect of structural invariants and is called the *flatband voltage*.

We will rely on (3.29) to analyze the conditions of the surface space-charge region. However, it needs a companion equation relating $\mathscr{E}(0)$ to ψ_s. This second equation comes from an integral of Poisson's equation. Writing

$$\frac{d^2\psi}{dx^2} = -\frac{q}{\epsilon_s}(p - n - N_A)$$

for the substrate, substituting (1.105) and (1.106) for n and p, deriving the equations

$$\phi_{Fn} = V + \phi_{FB} - \psi \tag{3.31}$$

$$\phi_{Fp} = \phi_{FB} - \psi \tag{3.32}$$

from the substrate energy-band diagram redrawn in Figure 3.4 for an arbitrary ψ_s, and using them for ϕ_{Fn} and ϕ_{Fp}, we obtain

$$\frac{d^2\psi}{dx^2} = F(\psi) \tag{3.33}$$

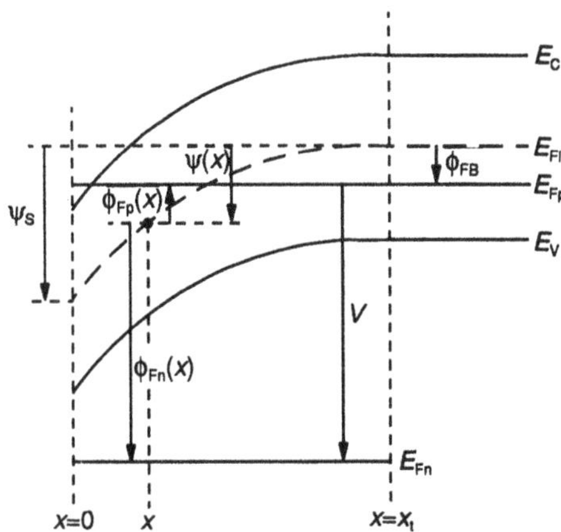

Figure 3.4 Details of the transversal energy-band diagram in the substrate.

where

$$F(\psi) \equiv -\frac{q}{\epsilon_s}\left\{n_i \exp\left[\frac{q}{kT}(\phi_{FB} - \psi)\right] - n_i \exp\left[-\frac{q}{kT}(\phi_{FB} - \psi + V)\right] - N_A\right\}$$

Multiplying both sides of (3.33) by $2d\psi/dx$, and recognizing the equation

$$2\frac{d\psi}{dx}\left(\frac{d^2\psi}{dx^2}\right) = \frac{d}{dx}\left(\frac{d\psi}{dx}\right)^2$$

we can transform (3.33) into

$$d\left(\frac{d\psi}{dx}\right)^2 = 2F(\psi)\, d\psi$$

Integrating this equation with the boundary conditions

$$\frac{d\psi}{dx}\Big|_{x=x_t} = 0, \quad \frac{d\psi}{dx}\Big|_{x=0} = -\mathscr{E}(0), \quad \psi|_{x=x_t} = 0, \quad \psi|_{x=0} = \psi_s$$

we finally obtain the desired equation

$$\mathscr{E}(0) = \pm\left(\frac{2qn_i}{\epsilon_s}\frac{kT}{q}\left\{\exp\left[\frac{q}{kT}(\phi_{FB} - \psi_s)\right] - \exp\left(\frac{q}{kT}\phi_{FB}\right) + \exp\left[-\frac{q}{kT}(\phi_{FB} + V - \psi_s)\right] - \exp\left[-\frac{q}{kT}(\phi_{FB} + V)\right] + \frac{q}{kT}\frac{N_A}{n_i}\psi_s\right\}\right)^{1/2} \tag{3.34}$$

Having (3.29) and (3.34) in hand, we can now proceed to an analysis of the five surface conditions of flatband, threshold, accumulation, depletion, and inversion.

Flatband

As previously defined, flatband is the surface condition in which the surface concentration of the majority carriers equals the doping concentration, that is $p(0) = N_A$ in an n-channel MOSFET. Since holes have the same concentration in the bulk, this condition implies a position-independent ϕ_{Fp} equaling ϕ_{FB}, hence, no band bending and therefore no space charge, as shown in Figure 3.5. Since ψ_s is zero, (3.34) indicates $\mathscr{E}(0) = 0$, and thus, (3.29) yields

$$V_G = V_{FB} \quad \text{(flatband condition)} \tag{3.35}$$

It is obvious that, as long as the gate voltage satisfies (3.35), a flatband condition prevails regardless of the value of V and, therefore, of the lateral position y.

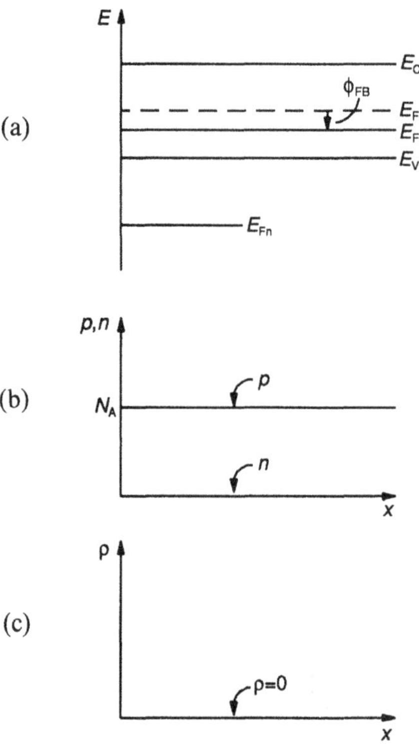

Figure 3.5 Internal variables at flatband: (a) energy-band diagram, (b) carrier concentration profiles, and (c) charge density profile.

Threshold

Threshold, being the condition in which the minority-carrier surface concentration has the same value as the doping concentration, implies the equation

$$n(0) = N_A \tag{3.36}$$

for an n-channel MOSFET. Since

$$n(0) = n_i \exp\left[-\frac{q}{kT}\phi_{Fn}(0)\right] = n_i \exp\left[-\frac{q}{kT}(V + \phi_{FB} - \psi_s)\right] \tag{3.37}$$

and

$$N_A = n_i \exp\left(\frac{q}{kT}\phi_{FB}\right) \tag{3.38}$$

(3.36) yields

$$\psi_s = V + 2\phi_{FB} \tag{3.39}$$

which corresponds to the energy-band diagram given in Figure 3.6(a). It is obvious from this diagram that, in the surface space-charge region, n is a decreasing function of x varying between $n(0) = N_A$ and $n(x_t) = (n_i^2/N_A)\exp(-qV/kT)$, whereas p is an increasing function of x varying between $p(0) = (n_i^2/N_A)\exp(-qV/kT)$ and $p(x_t) = N_A$, as shown in Figure 3.6(b). The resulting charge distribution in the space-charge region, shown in Figure 3.6(c), is virtually due to the acceptor ions except for the very near vicinity of the surface where the electron charge density is comparable to the dopant charge density. In other words, most of the space-charge region is in a depletion mode.

The boundary field under the threshold condition can be obtained from (3.34) simply by substituting (3.39) into the latter:

$$\mathscr{E}(0) = \sqrt{\frac{2qN_A}{\epsilon_s}(V + 2\phi_{FB})} \tag{3.40}$$

Notice the selection of the positive solution from (3.34). This is due to the fact that the energy bands increase with position, thus indicating a positive field in the space-charge region including the boundary.

Substituting (3.40) together with (3.39) into (3.29), we can specify the gate voltage that induces the threshold condition:

$$V_G = V_T(V) + V \quad \text{(threshold condition)} \tag{3.41}$$

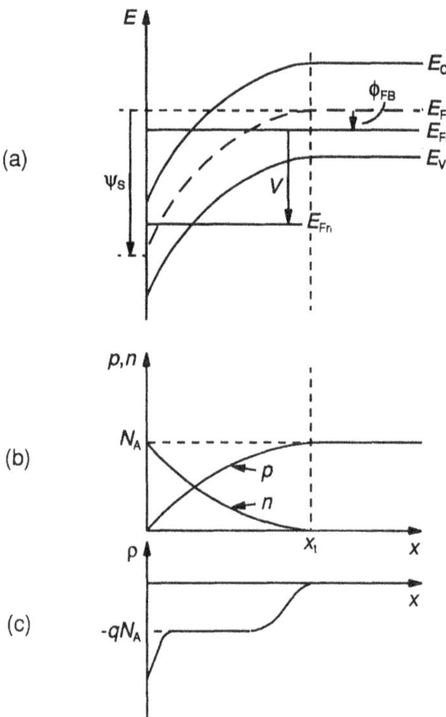

Figure 3.6 Internal variables at threshold: (a) energy-band diagram, (b) carrier concentration profiles, and (c) charge density profile.

where

$$V_T(V) \equiv V_{FB} + 2\phi_{FB} + \gamma\sqrt{V + 2\phi_{FB}} \tag{3.42}$$

and

$$\gamma \equiv \frac{1}{C_{ox}}\sqrt{2q\epsilon_s N_A} \tag{3.43}$$

are called the *threshold voltage* and *body factor*, respectively.

Accumulation

In accumulation, the majority-carrier concentration at the surface must exceed the doping concentration, which means

$$p(0) > N_A$$

for an n-channel MOSFET. Substituting into this inequality (3.38) for N_A and

$$p(0) = n_i \exp\left[\frac{q}{kT}\phi_{Fp}(0)\right] = n_i \exp\left[\frac{q}{kT}(\phi_{FB} - \psi_s)\right]$$

for $p(0)$, we obtain $\psi_s < 0$, which implies an energy-band diagram like that of Figure 3.7(a). Since accumulation is of little, if any, significance in MOSFET current-voltage modeling, we do not carry its analysis any further than this. But the reader can easily show that a positive charge density distribution develops in the surface space-charge region due to the dominance of holes as shown in Figure 3.7(b), and that in order to induce accumulation, the boundary field must be negative and V_G must be less than V_{FB}.

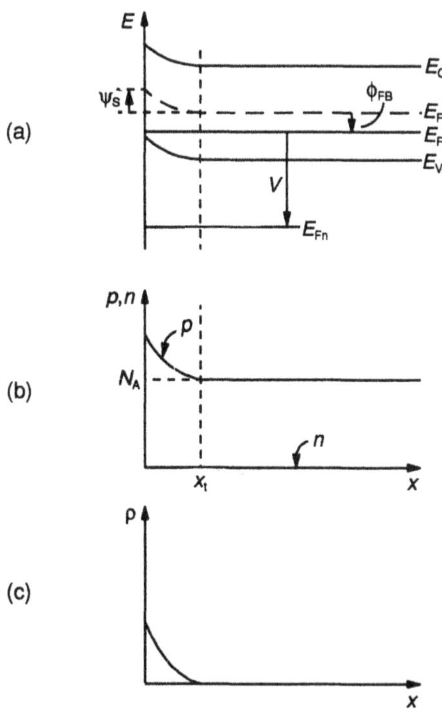

Figure 3.7 Internal variables at accumulation: (a) energy-band diagram, (b) carrier concentration profiles, and (c) charge density profile.

Depletion

We know that the surface concentration of holes in flatband and of electrons in threshold equals the doping concentration. The depletion condition, which represents the case of both carriers having surface concentrations less than the doping concentration, should therefore fit between the flatband and threshold conditions and be characterized by

$$0 < \psi_s < V + 2\phi_{FB}$$

$$0 < \mathscr{E}(0) < \sqrt{\frac{2qN_A}{\epsilon_s}(V + 2\phi_{FB})}$$

and

$$V_{FB} < V_G < V_T(V) + V$$

in an n-channel MOSFET. A typical energy-band diagram, carrier concentration profiles, and a charge density profile are shown in Figure 3.8(a–c), respectively. Since both types of carriers have smaller concentrations in comparison with the doping concentration, the space-charge region is characterized by a position-in-

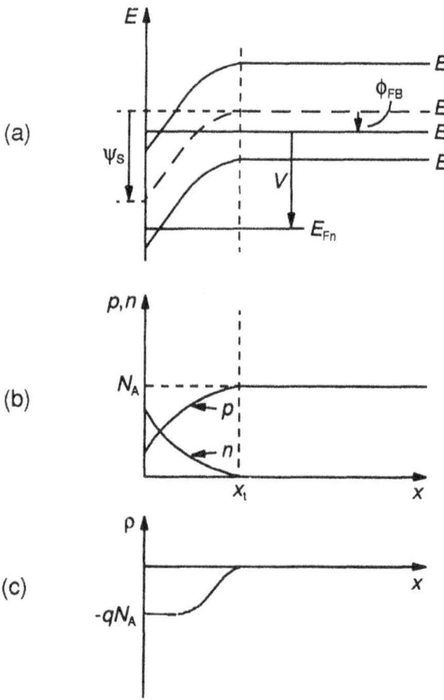

Figure 3.8 Internal variables at depletion: (a) energy-band diagram, (b) carrier concentration profiles, and (c) charge density profile.

dependent charge density profile of $\rho = -qN_A$. Integrating Poisson's equation for this profile, we obtain the following relationship between x_t and $\mathscr{E}(0)$:

$$\mathscr{E}(0) = \frac{qN_A}{\epsilon_s}x_t \tag{3.44}$$

In addition, we can write

$$\psi_s = \frac{\epsilon_s}{2qN_A}\mathscr{E}^2(0) \tag{3.45}$$

either from a second integration of Poisson's equation, or alternatively from (3.34), in which only the last term in the braces is significant under depletion conditions. Substituting (3.45) into (3.29) for ψ_s, replacing $\mathscr{E}(0)$ with the right-hand side of (3.44), and solving for x_t, we can relate the thickness of the depleted surface space-charge region to the gate voltage as

$$x_t = -\frac{\epsilon_s}{C_{ox}} + \sqrt{\left(\frac{\epsilon_s}{C_{ox}}\right)^2 + \frac{2\epsilon_s}{qN_A}(V_G - V_{FB})} \tag{3.46}$$

which indicates x_t as an increasing function of V_G. For a physical interpretation of this result, it suffices to consider the facts: (1) The boundary field increases with the gate voltage, (2) an increasingly positive boundary field attracts an increasing amount of negative charge to the surface, and (3) under depletion conditions, in which carrier concentrations are negligible, the negative surface charge can increase only by an exposure of more acceptor ions, which calls for an expansion of the depletion region boundary.

Inversion

An *n*-channel MOSFET is inverted when the surface concentration of electrons exceeds the doping concentration. This condition implies

$$\psi_s > V + 2\phi_{FB}$$

$$\mathscr{E}(0) > \sqrt{\frac{2qN_A}{\epsilon_s}(V + 2\phi_{FB})}$$

and

$$V_G > V_T(V) + V$$

Shown in Figure 3.9(a) is a typical energy-band diagram in inversion. The corresponding carrier concentration and charge-density profiles are given in Figure

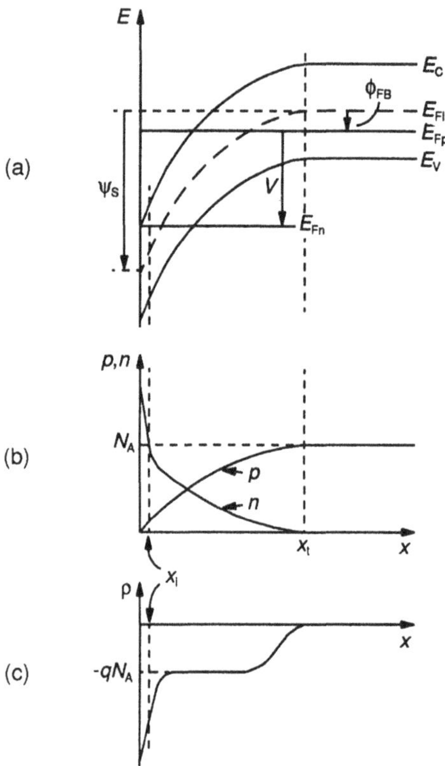

Figure 3.9 Internal variables at inversion: (a) energy-band diagram, (b) carrier concentration profiles, and (c) charge density profile.

3.9(b–c), respectively. The narrow region $0 \leq x \leq x_i$, in which n exceeds N_A, is called the *inversion layer*. Next to this layer, we observe a depletion layer extending between $x = x_i$ and $x = x_t$ and supporting an electrostatic potential difference $V + 2\phi_{FB}$. The depletion thickness corresponding to this potential difference can be expressed (by integrating Poisson's equation) as follows:

$$x_t - x_i = \sqrt{\frac{2\epsilon_s}{qN_A}(V + 2\phi_{FB})} \tag{3.47}$$

which, apparently, is independent of V_G. The depletion region attains this thickness at the threshold and remains constant for larger values of V_G. This means that the charge contributed by acceptor ions is constant in inversion. However, an increasing

V_G enhances the boundary field and, therefore, necessitates more negative charge in the surface space-charge region, which, obviously, is supplied by the electrons forming the inversion layer. Note that however large the charge residing in the inversion layer may be, the thickness x_i of this layer is generally a very small fraction of the thickness of the space-charge layer. For this reason, the latter virtually equals the depletion thickness given by (3.47).

Another inversion property of significance is the reduction of the sensitivity of ψ_s to V_G. The reader can easily show from (3.29) and (3.45) that, in depletion, ψ_s is an approximately linearly increasing function of V_G. As the device enters inversion, the rate of change in ψ_s with V_G starts slowing down, and under extremely strong inversion conditions it reduces to a very slow logarithmic rate, as one can show from (3.29) and (3.34). We therefore conclude that ψ_s cannot increase much beyond its threshold value $V + 2\phi_{FB}$.

3.2.3 A General Strong Inversion Model

In Section 3.2.1, after deriving (3.19), we pointed out that three more analytical steps had to be taken to complete the development of a MOSFET model. Relating the surface electron charge density Q_n to the channel potential V is the first of these steps. Ideally, we would like to do this in a unified manner for both inversion and depletion conditions including the threshold. In this manner, we arrive at a single model describing all levels of MOSFET current. Unfortunately, the fact that no analytical solution is available for such a unified treatment forces us to develop two separate models for these two cases. The MOSFET operation under depletion conditions is described by what is most commonly known as the *subthreshold* model, whose derivation is presented in Section 3.2.5. In the remainder of this section we will develop the so-called *strong inversion* model to describe MOSFET behavior mainly for the inversion condition, hence, for relatively high levels of channel current. The derivation will closely follow that of Meyer [3]. Simplified versions of this model are discussed in Section 3.2.4.

For an inverted surface space-charge region, Poisson's equation can be expressed only in terms of n and N_A as

$$\frac{d\mathscr{E}_x}{dx} = \frac{q}{\epsilon_s}(-n - N_A)$$

because p is negligible in the region. Integrating this equation between $x = 0$ and $x = x_t$, assuming $\mathscr{E}_x(x_t) = 0$ due to the quasineutrality of the substrate, and recalling (3.16), we obtain

$$Q_n = \epsilon_s\mathscr{E}(0) - qN_Ax_t \tag{3.48}$$

We know from the preceding section that x_t is virtually equal to the thickness of the depletion region. This knowledge enables us to substitute the right-hand side of (3.47) for x_t in (3.48). For $\mathscr{E}(0)$, we can use (3.29), in which $V + 2\phi_{FB}$ can be substituted for ψ_s because the latter cannot be much greater than its threshold level, as discussed at the end of the preceding section. These substitutions turn (3.48) into

$$Q_n = C_{ox}[V_G - V_{FB} - 2\phi_{FB} - V - \gamma(V + 2\phi_{FB})^{1/2}] \tag{3.49}$$

which also can be written with the aid of (3.42) as

$$Q_n = C_{ox}[V_G - V - V_T(V)] \tag{3.50}$$

We now briefly assess this expression before using it in the final phase of model development. First, note that a substitution of (3.41) into (3.50) yields $Q_n = 0$ for the threshold condition. However, this result does not accurately reflect the actual physical situation because, although no inversion layer is present under the threshold condition, a nonzero, but small, electron population still exists in the space-charge region. This population is not represented in (3.49) and (3.50) because the assumption of a "depleted" space-charge layer made in the derivation of (3.47) removed it from the analytical scene. In other words, Q_n, as expressed by (3.49) and (3.50) is assumed to be due only to those electrons that reside inside an inversion layer. This assumption obviously impairs the accuracy of these equations under threshold and *weak inversion* conditions, in which the contribution of inversion-layer electrons to Q_n is less than or comparable to that of depletion-layer electrons. However, the accuracy increases rapidly with the increasing strength of inversion. This is why the model we are about to present is essentially a strong inversion model.

Equations (3.49) and (3.50) are valid at any y provided that a strong inversion condition prevails. We now identify three possible cases regarding the extent of strong inversion along y, and will consider each of these cases separately in the final phase of model development.

The first case to be considered is characterized by a strong inversion condition prevailing all the way between source and drain. The operation mode corresponding to this physical situation is called *nonsaturation*. The second case involves a lack of inversion at the drain side of the surface space-charge region and is called *saturation*. The final case is the one in which no part of the surface space-charge region is inverted. This corresponds to the so-called *cutoff* mode of operation.

Nonsaturation

Now suppose that a strongly inverted channel is present everywhere between the source and drain junctions, as shown in Figure 3.10(a), so that (3.49) is valid at

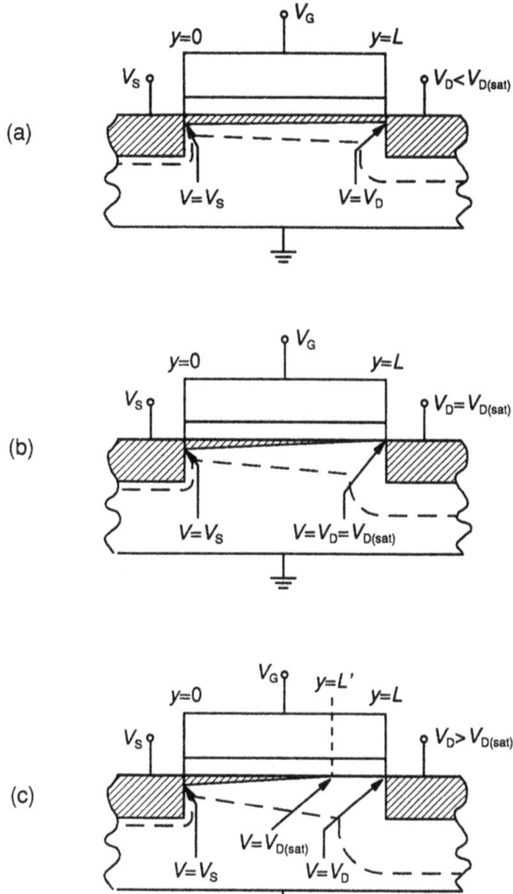

Figure 3.10 The formation of a strong-inversion layer in (a) nonsaturation, (b) onset of saturation, and (c) saturation.

all transversal locations $0 \leq y \leq L$. Substituting (3.49) for Q_n in (3.19), and integrating between $y = 0$ and $y = L$, we obtain

$$I\int_0^L dy = W\overline{\mu}_n C_{ox}\int_{V|_{y=0}}^{V|_{y=L}} [V_G - V_{FB} - 2\phi_{FB} - V - \gamma(V + 2\phi_{FB})^{1/2}]\, dV \quad (3.51)$$

Using the boundary conditions (3.23) and (3.24) and integrating, we finally turn (3.51) into

$$I = K_P\frac{W}{L}\left\{\left[V_G - V_{FB} - 2\phi_{FB} - \frac{1}{2}(V_D + V_S)\right](V_D - V_S) - \frac{2}{3}\gamma[(V_D + 2\phi_{FB})^{3/2} - (V_S + 2\phi_{FB})^{3/2}]\right\} \tag{3.52}$$

where the so-called transconductance parameter K_P is defined by $K_P \equiv \overline{\mu}_n C_{ox}$. In some applications, it is more convenient to use the following alternative form of this equation:

$$I = K_P\frac{W}{L}\left\{\left(V_{GS} - V_{FB} - 2\phi_{FB} - \frac{1}{2}V_{DS}\right)V_{DS} - \frac{2}{3}\gamma[(V_{DS} + V_S + 2\phi_{FB})^{3/2} - (V_S + 2\phi_{FB})^{3/2}]\right\} \tag{3.53}$$

where $V_{GS} \equiv V_G - V_S$ and $V_{DS} \equiv V_D - V_S$ are the source-referred gate and drain voltages, respectively.

Equations (3.52) and (3.53) are two alternative characteristic equations describing MOSFET behavior when an inversion channel bridges the source and drain junctions. This operation mode is called *nonsaturation*. We now determine the necessary and sufficient condition of this mode in terms of bias voltages. Since strong inversion conditions prevail at both $y = 0$ and $y = L$, the conditions $Q_n|_{V=V_S} \geq 0$ and $Q_n|_{V=V_D} \geq 0$ must be satisfied. Using (3.50), we can transform these conditions into

$$V_G - V_S \geq V_T \quad \text{strong inversion at source} \tag{3.54}$$

and

$$V_G - V_D \geq V_T' \quad \text{strong inversion at drain} \tag{3.55}$$

where V_T is defined by

$$V_T \equiv V_T(V_S) = V_{FB} + 2\phi_{FB} + \gamma(V_S + 2\phi_{FB})^{1/2} \tag{3.56}$$

and is called the *MOSFET threshold voltage* and V_T' is defined by

$$V_T' \equiv V_T(V_D) = V_{FB} + 2\phi_{FB} + \gamma(V_D + 2\phi_{FB})^{1/2} \tag{3.57}$$

Since $V_D \geq V_S$ and, therefore, $V_T' \geq V_T$, (3.55) is a more stringent condition compared to (3.54). Hence, (3.54) is redundant and (3.55) is the necessary and sufficient condition of nonsaturation. It is possible, and in some cases preferable, to express (3.55) in terms of the source-referred bias voltages as follows

$$V_{GS} - V_T \geq V_{DS} + \Delta V_T \tag{3.58}$$

where

$$\Delta V_T \equiv V_T' - V_T = \gamma[(V_D + 2\phi_{FB})^{1/2} - (V_S + 2\phi_{FB})^{1/2}] \tag{3.59}$$

We therefore conclude that, if condition (3.55) or, alternatively, (3.58) is satisfied, the device will operate in nonsaturation and will conduct the current described by (3.52) or (3.53).

Saturation

Since (3.55) is more stringent than (3.54), we must also recognize the following possibility:

$$V_G - V_S \geq V_T \quad \text{strong inversion at the source end,} \tag{3.60}$$

$$V_G - V_D < V_T' \quad \text{no inversion at the drain end.} \tag{3.61}$$

In terms of source-referred bias voltages, these conditions can be stated as

$$V_{DS} + \Delta V_T > V_{GS} - V_T \geq 0 \tag{3.62}$$

This mode of MOSFET operation is called *saturation*.

Since the drain end is not inverted in saturation, (3.52) and (3.53) can no longer describe the relationship between the port variables. It is obvious from (3.60) and (3.61), however, that a strong inversion condition still prevails between the source end and an intermediate point $y = L'$ as shown in Figure 3.10(c). For this reason, the inversion charge equation (3.49) is still valid for the region $0 \leq y \leq L'$. Substituting (3.49) into (3.19) and integrating the latter over this region, we obtain

$$I \int_0^{L'} dy = W\overline{\mu}_n C_{ox} \int_{V_S}^{V|_{y=L'}} [V_G - V_{FB} - 2\phi_{FB} - V - \gamma(V + 2\phi_{FB})^{1/2}]\, dV \tag{3.63}$$

which differs from (3.51) only in the upper limit of the integral. Therefore, (3.63) should yield an expression for I in the same form as (3.52) with the exception that L is replaced by L' and V_D by $V|_{y=L'}$. Since $y = L'$ is defined as the point where Q_n vanishes, we can write the following expression for $V|_{y=L'}$ simply by equating the right-hand side of (3.49) to zero:

$$V|_{y=L'} = V_{D(\text{sat})} \tag{3.64}$$

where

$$V_{D(\text{sat})} \equiv V_G - V_{FB} - 2\phi_{FB} + \frac{\gamma^2}{2} - \gamma\sqrt{V_G - V_{FB} + \frac{\gamma^2}{4}} \tag{3.65}$$

Rewriting (3.52) after replacing L with L' and V_D with $V_{D(\text{sat})}$ we obtain the following characteristic equation of the saturation mode of operation:

$$\begin{aligned} I = K_P\frac{W}{L'}\Bigg(&\left\{V_G - V_{FB} - 2\phi_{FB} - \frac{1}{2}\left[V_{D(\text{sat})} + V_S\right]\right\}(V_{D(\text{sat})} - V_S) \\ &- \frac{2}{3}\gamma[(V_D + 2\phi_{FB})^{3/2} - (V_S + 2\phi_{FB})^{3/2}]\Bigg) \end{aligned} \tag{3.66}$$

which can also be expressed as

$$\begin{aligned} I = K_P\frac{W}{L'}\Bigg\{&\left[V_{GS} - V_{FB} - 2\phi_{FB} - \frac{1}{2}V_{DS(\text{sat})}\right]V_{DS(\text{sat})} \\ &- \frac{2}{3}\gamma\,[(V_{DS(\text{sat})} + V_S + 2\phi_{FB})^{3/2} - (V_S + 2\phi_{FB})^{3/2}]\Bigg\} \end{aligned} \tag{3.67}$$

where $V_{DS(\text{sat})} \equiv V_{D(\text{sat})} - V_S$.

In the derivation (3.66) or (3.67) of the characteristic equation of saturation, we relied solely on the strongly inverted channel section. Since the current is spatially constant in a dc steady state, the noninverted section extending between $y = L'$ and $y = L$ must be capable of supporting the very same current. How can a section of vanishingly small electron concentration conduct a current imposed on it by the channel? The first step toward a convincing answer is to consider the case in which the device operates just at the boundary between the saturation and nonsaturation modes. According to (3.55) and (3.61), this boundary operation occurs when $V_D = V_G - V_T'$, which, obviously, is the threshold condition for the drain end. Substituting (3.57) for V_T' in this equation and solving for V_D, we obtain $V_D = V_{D(\text{sat})}$ for a drain voltage that biases the device to this boundary mode. The corresponding physical situation is illustrated in Figure 3.10(b). The endpoint of the inverted section is coincident with $y = L$ where the potential is $V_{D(\text{sat})}$. Now suppose V_D is increased above $V_{D(\text{sat})}$ so that the device enters saturation as shown in Figure 3.10(c). We know from (3.64) that the potential at the endpoint of the strongly inverted section, $V|_{y=L'}$, must remain at $V_{D(\text{sat})}$. But since V_D is now greater than $V_{D(\text{sat})}$, a positive potential difference will develop between the edge of the drain junction and the endpoint of the inverted section. Since this voltage cannot drop in an infinitesimally small distance, the endpoint has to move away from the drain edge. This is how a noninverted *pinch-off* section is created. The voltage across this section obviously results in a lateral electric field that is directed toward the endpoint. Those electrons arriving from the source along the inverted section and reaching the endpoint are accelerated by this field in the pinch-off region and subsequently collected by the drain. This conduction process resembles the passage of electrons through the reverse-biased CB depletion region of a BJT operating in the forward active mode. In both cases, the region is depleted, but the field is so large that a large amount of drift current can be conducted by a small concentration of fast electrons. Also note that the electric-field pattern in the pinch-off region is two dimensional due to the proximity of the drain junction. An accurate quantitative description is difficult, but certainly the field must be distributed in such a way that the continuity of the current is maintained and the two-dimensional Poisson's equation is satisfied. The location of the endpoint $y = L'$ is determined by these constraints. Qualitatively, we expect L' to be a decreasing function of V_D because the larger the latter, the wider the noninverted section to absorb it. As understood from (3.66) or (3.67), L' is the only parameter that can make the current V_D dependent in saturation. To be specific, the current becomes an increasing function of V_D due to the shrinking channel, which is commonly known as the *channel-length modulation effect*. In a long-channel device, whose L is large, this effect is insignificant, and L' can be replaced with L without much error. In short-channel devices, on the other hand, the noninverted section can be a very significant part of the channel length and, therefore, can introduce a considerable V_D dependence to the current. We return to this subject in Section 3.4.2.

Cutoff

The third, and final, bias condition is called *cutoff*, and is defined as follows:

$$V_G - V_S < V_T \quad \text{no inversion at the source end} \tag{3.68}$$

$$V_G - V_D < V_T' \quad \text{no inversion at the drain end} \tag{3.69}$$

Since $V_D \geq V_S$, and therefore $V_T' \geq V_T$, the condition stated by (3.69) is redundant, and (3.68) is the necessary and sufficient condition of cutoff. The latter can be stated also as

$$V_{GS} - V_T < 0 \tag{3.70}$$

Since no channel exists in cutoff, the strong inversion model assumes $Q_n = 0$ everywhere. This implies

$$I = 0 \tag{3.71}$$

in cutoff. In reality, a small subthreshold current can flow, but its modeling requires a separate approach, as mentioned previously.

3.2.4 Simplified Strong Inversion Models

The model developed in the preceding section is fairly accurate. Unfortunately, however, the equations involved are too complicated to be suitable for manual circuit analysis. We will now reduce their complexity in two successive steps. This will lead us first to an intermediate model of moderate simplicity and accuracy [4]. A further simplification will result in a lower order model of reduced accuracy but of higher utility [5].

An Intermediate Model

The first step of simplification involves the term $(V_{DS} + V_S + 2\phi_{FB})^{3/2}$ appearing in (3.53). Rewriting it as

$$(V_S + 2\phi_{FB})^{3/2}\left(1 + \frac{V_{DS}}{V_S + 2\phi_{FB}}\right)^{3/2}$$

and expanding the latter into Taylor's series around $V_{DS} = 0$, we obtain

$$(V_S + 2\phi_{FB})^{3/2}\left[1 + \frac{3}{2}\frac{V_{DS}}{V_S + 2\phi_{FB}} + \frac{3}{8}\frac{V_{DS}^2}{(V_S + 2\phi_{FB})^2} + \ldots\right] \quad (3.72)$$

Ignoring the terms of order 3 and higher, and substituting (3.72) back into (3.53), we obtain

$$I = K_P\frac{W}{L}\left[V_{GS} - V_T - \frac{1}{2}(1 + \delta')V_{DS}\right]V_{DS} \quad (3.73)$$

as the model equation for nonsaturation. Here, the threshold voltage V_T is still given by (3.56), and δ' is defined by

$$\delta' \equiv \frac{\gamma}{2(V_S + 2\phi_{FB})^{1/2}} \quad (3.74)$$

The necessary and sufficient condition of nonsaturation (3.58) is now simplified into

$$V_{GS} - V_T \geq (1 + \delta')V_{DS} \quad (3.75)$$

which can be obtained by expanding (3.59) into a Taylor's series around $V_{DS} = 0$ and selecting the first two terms. Similarly, the condition of saturation (3.62) will become

$$(1 + \delta')V_{DS} > V_{GS} - V_T \geq 0 \tag{3.76}$$

To simplify the modeling equation of saturation, we can rewrite the nonsaturation equation (3.73) for the onset of saturation by substituting $V_{DS(\text{sat})}$ for V_{DS}. Since $V_{DS(\text{sat})}$ is the drain-source voltage for which the transition from nonsaturation to saturation occurs, we can write from (3.75) the following equation:

$$V_{DS(\text{sat})} = \frac{V_{GS} - V_T}{1 + \delta'} \tag{3.77}$$

which, in turn, reduces (3.73) into

$$I = \frac{1}{2} K_P \frac{W}{L'} \frac{1}{1 + \delta'} (V_{GS} - V_T)^2 \tag{3.78}$$

as the simplified model equation for saturation.

A Simple Model

An ultimately simple model can be obtained by repeating the above derivation with only the first two terms of (3.72). This will result in

$$I = K_P \frac{W}{L} \left(V_{GS} - V_T - \frac{1}{2} V_{DS}\right) V_{DS} \tag{3.79}$$

and

$$V_{GS} - V_T \geq V_{DS} \tag{3.80}$$

for nonsaturation, and

$$I = \frac{1}{2} K_P \frac{W}{L'} (V_{GS} - V_T)^2 \tag{3.81}$$

and

$$V_{DS} > V_{GS} - V_T \geq 0 \tag{3.82}$$

for saturation.

Accuracy of the Simplified Models

Since the simplified models are based on a truncated series expansion of (3.53) around $V_{DS} = 0$, their accuracy is relatively high for small values of V_{DS}. This is illustrated in Figure 3.11, where the output characteristics of a typical MOSFET as predicted by the accurate model as well as the simplified models are given. The accuracy is expected to improve also with an increasing V_S because the larger the latter, the smaller the truncation error in (3.72). Finally, the body factor γ, being a prefactor of the series-approximated part of (3.53), determines the effect of the truncation error on I. Obviously, the smaller the γ, the better the accuracy.

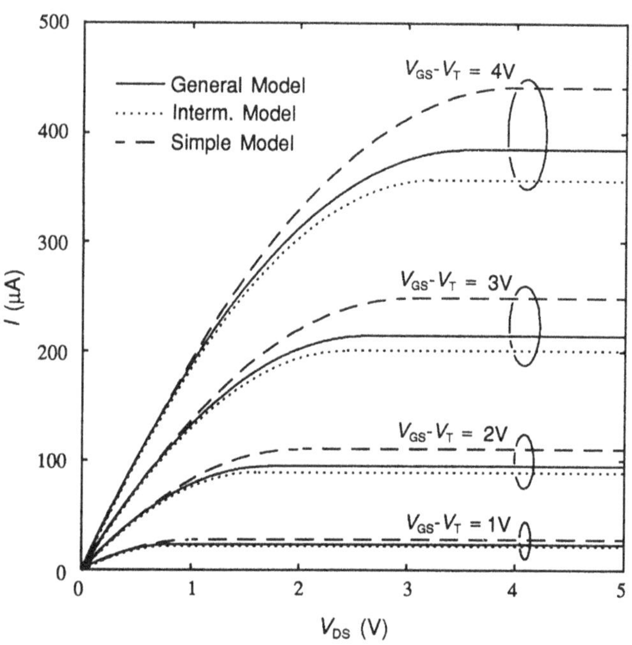

Figure 3.11 A comparison of MOSFET output characteristics as predicted by the accurate, intermediate, and simple strong inversion models. The MOSFET is specified with $N_A = 2 \times 10^{15}$ cm^{-3}, $T_{ox} = 50$ nm, $Q_{ss} = 0$, $\phi_{FG} = 0.55$V, $W/L = W/L' = 1$, and $V_S = 0$.

3.2.5 A Subthreshold Model

An n-channel MOSFET operating in the bias range $V_{FB} < V_G < V_S + V_T$ has a depleted surface space-charge region. Although strong inversion models predict $I = 0$ for this bias range, a sufficiently large Q_n and, therefore, I can still be generated especially at the high end of the range. In modeling this *subthreshold* current, we cannot rely on (3.49) or (3.50) for representing the relationship between Q_n and V because these equations are valid only under strong inversion conditions. Instead, we need to determine the transversal electron concentration profile $n(x)$ first, and then integrate this profile according to (3.16) to obtain the desired relationship.

Electron concentration at a given location x is described quite generally by $n(x) = n_i \exp[-q\phi_{Fn}(x)/kT]$, and $\phi_{Fn}(x)$ is given by (3.31). Hence,

$$n(x) = n_i \exp\left\{\frac{q}{kT}[\psi(x) - \phi_{FB} - V]\right\} \tag{3.83}$$

The profile $\psi(x)$ of the electrostatic potential can be determined from Poisson's equation assuming depletion, that is,

$$\frac{d^2\psi(x)}{dx^2} = \frac{q}{\varepsilon_s}N_A$$

Integrating this equation with the boundary conditions $\psi(x_t) = 0$, $d\psi/dx|_{x=x_t} = -\mathscr{E}(x_t) = 0$ and $d\psi/dx|_{x=0} = -\mathscr{E}(0)$ we obtain

$$\psi(x) = \frac{\epsilon_s}{2qN_A}\mathscr{E}^2(0) - \mathscr{E}(0)x + \frac{qN_A}{2\epsilon_s}x^2 \tag{3.84}$$

Under depletion conditions, the surface transversal field $\mathscr{E}(0)$ is positive; therefore, $\psi(x)$ is a decreasing function of x, which, according to (3.83), implies that Q_n is mostly due to those electrons located very close to the interface. On the other hand, the last term on the right-hand side of (3.84) is negligibly small in the near vicinity of the interface especially when the device operates close to inversion with a relatively large $\mathscr{E}(0)$. Ignoring this term, and substituting the remainder of (3.84) into (3.83), we obtain

$$n(x) \cong n_i \exp\left[\frac{q}{kT}\left(\frac{\epsilon_s}{2qN_A}\mathscr{E}^2(0) - \phi_{FB} - V\right)\right]\exp\left[-\frac{q}{kT}\mathscr{E}(0)x\right] \tag{3.85}$$

which clearly shows the exponential decaying of $n(x)$ with position. We can now obtain Q_n by integrating (3.85) in accordance with (3.16). To simplify the result, the upper limit $x = x_t$ can be replaced by $x = \infty$ because, in any case, $n(x)$ becomes negligible well before the boundary $x = x_t$ is reached. The result follows:

$$Q_n = \frac{kT}{q}\frac{n_i^2}{N_A}\frac{q}{\mathscr{E}(0)}\exp\left[\frac{q}{kT}\frac{\epsilon_s}{2qN_A}\mathscr{E}^2(0)\right]\exp\left(-\frac{q}{kT}V\right) \tag{3.86}$$

Next, we relate the boundary field to the gate voltage by canceling x_t between (3.44) and (3.46), that is,

$$\mathscr{E}(0) = \frac{qN_A}{C_{ox}}\left[\sqrt{1 + \frac{4}{\gamma^2}(V_G - V_{FB})} - 1\right] \tag{3.87}$$

Finally, substituting (3.87) into (3.86), using the latter in (3.19), and integrating from source to drain with the boundary conditions $V|_{y=0} = V_S$, $V|_{y=L} = V_D$, we obtain the following subthreshold characteristic equation of an n-channel MOSFET:

$$I = K_P \frac{W}{L} \frac{(kTn_i/qN_A)^2}{\sqrt{1 + (4/\gamma^2)(V_G - V_{FB})} - 1} \exp\left\{\frac{q}{kT}\frac{\gamma^2}{4} \left[\sqrt{1 + \frac{4}{\gamma^2}(V_G - V_{FB})} - 1\right]^2\right\} \left[\exp\left(-\frac{q}{kT}V_S\right) - \exp\left(-\frac{q}{kT}V_D\right)\right] \tag{3.88}$$

An alternative form, expressed in terms of the source-referred bias voltages, V_{GS} and V_{DS}, is as follows:

$$I = \frac{1}{2} K_P \frac{W}{L}\left(\frac{kT}{q}\frac{n_i}{N_A}\gamma\right)^2 \frac{\exp[(q/kT)(V_{GS} - V_{TX})]}{V_{TX} - V_{FB}} \left[1 - \exp\left(-\frac{q}{kT}V_{DS}\right)\right] \tag{3.89}$$

where V_{TX} is defined by

$$V_{TX} \equiv V_{FB} - \frac{\gamma^2}{2} + \gamma\sqrt{\frac{\gamma^2}{4} + V_{GS} - V_{FB} + V_S} \tag{3.90}$$

The subthreshold transfer characteristics, as predicted by (3.89) for the MOSFET specified in Figure 3.11, are shown in Figure 3.12. They clearly exhibit the exponential dependence of the current on V_{GS} resembling BJT Gummel plots. These characteristics also indicate a strong dependence on V_S but, according to (3.89), no dependence on V_{DS} is predicted provided that the latter exceeds a few kT/q. Another subthreshold property worth mentioning is the rather strong temperature sensitivity especially through the second-power dependence on the intrinsic carrier concentration n_i.

Also shown in Figure 3.12 are the transfer characteristics as predicted by (3.67) of the strong-inversion model. The discontinuity between the two models around $V_{GS} = V_T$ is not surprising because the presence of electrons in the depletion region is ignored in both models when Poisson's equation is integrated. This makes $V_{GS} = V_T$ the condition of worst accuracy for both. The accuracy of the subthreshold model deteriorates also at small gate voltages close to V_{FB} because the hole concentration in the surface space-charge region is ignored in the Poisson's equation used in model derivation. However, this error is of no practical significance because the channel current is already negligible for such small gate voltages.

The model developed in this section excludes a number of secondary effects, which can significantly alter device behavior in subthreshold operation. Carrier trapping at the Si-SiO_2 interface and a spatial variation of V_{FB} and N_A are the most

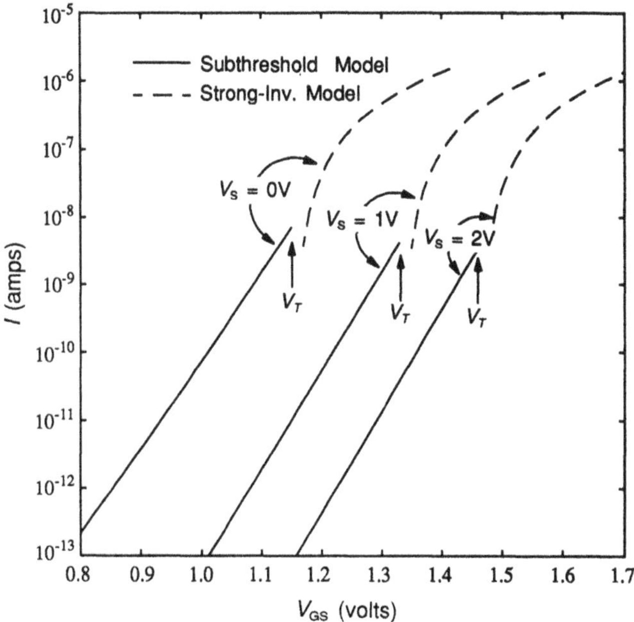

Figure 3.12 Subthreshold transfer characteristics for the MOSFET specified in Figure 3.11.

important ones. These effects cause the subthreshold slope of the logarithmic transfer characteristics to be less than that predicted by the primary model.

A large body of literature exists on the analysis and modeling of subthreshold conduction. For further reading on this matter, the reader is referred to Barron [6], Troutman [7, 8], Van Overstraeten et al. [9], and Brews [10].

3.2.6 *p*-Channel MOSFET

The *p*-channel operation can be analyzed and modeled along the same lines of the development presented in Sections 3.2.1 through 3.2.5. Of course, the roles of electrons and holes must be transposed. The key points to remember are the proper definition of the drain and source terminals [see (3.2)] and the imposition of nonforward biasing conditions on these terminals, that is, $V_D \leq 0$ and $V_S \leq 0$. Adopting the reference convention of Figures 3.1 and 3.2, writing $J_{py} = -J$, $J_{px} = 0$, $J_{ny} = 0$, $J_{nx} = 0$, and following the analysis presented in Section 3.2.1, we arrive at

$$I = W\overline{\mu}_p Q_p \frac{dV}{dy}$$

for the channel current. The channel charge density and potential are defined by

$$Q_p \equiv \int_0^{x_t} qp\,dx$$

$$V \equiv \phi_{Fp} - \phi_{Fn}$$

Analyzing the surface space-charge region of a p-channel MOSFET along the lines of Section 3.2.2 leads us to the very same equations derived in that section prior to (3.31). Equations (3.31) and (3.32), however, are modified as

$$\phi_{Fn} = \phi_{FB} - \psi$$

$$\phi_{Fp} = V + \phi_{FB} - \psi$$

and eventually lead us to

$$\mathscr{E}(0) = \pm\left(\frac{2qn_i}{\epsilon_s}\frac{kT}{q}\left\{\exp\left[\frac{q}{kT}(\psi_s - \phi_{FB})\right] - \exp\left(-\frac{q}{kT}\phi_{FB}\right) + \exp\left[\frac{q}{kT}(\phi_{FB} + V - \psi_s)\right]\right.\right.$$
$$\left.\left. - \exp\left[\frac{q}{kT}(\phi_{FB} + V)\right] - \frac{q}{kT}\frac{N_D}{n_i}\psi_s\right\}\right)^{1/2}$$

conditions (3.35) of flatband, (3.39) of surface potential at threshold, and (3.41) of gate voltage at threshold remain the same, but (3.40) of the boundary field at threshold, (3.42) of the threshold voltage, and (3.43) of the body factor become

$$\mathscr{E}(0) = -\sqrt{-\frac{2qN_D}{\epsilon_s}(V + 2\phi_{FB})}$$

$$V_T(V) \equiv V_{FB} + 2\phi_{FB} - \gamma\sqrt{-(V + 2\phi_{FB})}$$

and

$$\gamma \equiv \frac{1}{C_{ox}}\sqrt{2q\epsilon_s N_D}$$

respectively.

In a p-channel device, an accumulated surface space-charge region is characterized by an electron concentration in excess of the doping concentration, and occurs when $V_G > V_{FB}$. This also leads to a positive boundary field and a positive surface potential for accumulation.

A depletion condition prevails when $V_T(V) + V < V_G < V_{FB}$, for which ψ_s satisfies the condition $V + 2\phi_{FB} < \psi_s < 0$, and the boundary field is negative. An inversion condition is imposed when $V_G < V_T(V) + V$, which makes ψ_s more negative than $V + 2\phi_{FB}$.

The p-channel duals of the strong inversion model equations developed in Section 3.2.3 can be summarized as follows.

The nonsaturation mode of operation is subject to the condition

$$V_G - V_D \leq V_T'$$

or, alternatively

$$V_{GS} - V_T \leq V_{DS} - \Delta V_T$$

The channel current in this mode is described by

$$I = -K_P \frac{W}{L}\left\{\left[V_G - V_{FB} - 2\phi_{FB} - \frac{1}{2}(V_D + V_S)\right](V_D - V_S) + \frac{2}{3}\gamma\left[(-V_D - 2\phi_{FB})^{3/2} - (-V_S - 2\phi_{FB})^{3/2}\right]\right\}$$

or by

$$I = -K_P \frac{W}{L}\left\{\left(V_{GS} - V_{FB} - 2\phi_{FB} - \frac{1}{2}V_{DS}\right)V_{DS} + \frac{2}{3}\gamma\left[(-V_{DS} - V_S - 2\phi_{FB})^{3/2} - (-V_S - 2\phi_{FB})^{3/2}\right]\right\}$$

where

$$K_P \equiv \bar{\mu}_p C_{ox}$$

$$V_T' \equiv V_{FB} + 2\phi_{FB} - \gamma\sqrt{-(V_D + 2\phi_{FB})}$$

$$V_T \equiv V_{FB} + 2\phi_{FB} - \gamma\sqrt{-(V_S + 2\phi_{FB})}$$

and

$$\Delta V_T \equiv V_T' - V_T = \gamma[(-V_S - 2\phi_{FB})^{1/2} - (-V_D - 2\phi_{FB})^{1/2}]$$

The saturation mode of operation prevails under the conditions

$$V_G - V_S \leq V_T \quad \text{and} \quad V_G - V_D > V_T'$$

or alternatively,

$$V_{DS} - \Delta V_T < V_{GS} - V_T \leq 0$$

The channel current in saturation is given by

$$I = -\frac{W}{L'}K_P\left\{\left[V_G - V_{FB} - 2\phi_{FB} - \frac{1}{2}(V_{D(\text{sat})} + V_S)\right](V_{D(\text{sat})} - V_S) + \frac{2}{3}\gamma[(-V_{D(\text{sat})} - 2\phi_{FB})^{3/2} - (-V_S - 2\phi_{FB})^{3/2}]\right\}$$

or

$$I = -\frac{W}{L'}K_P\left\{\left(V_{GS} - V_{FB} - 2\phi_{FB} - \frac{1}{2}V_{DS(\text{sat})}\right)V_{DS(\text{sat})} + \frac{2}{3}\gamma[(-V_{DS(\text{sat})} - V_S - 2\phi_{FB})^{3/2} - (-V_S - 2\phi_{FB})^{3/2}]\right\}$$

where

$$V_{D(\text{sat})} = V_G - V_{FB} - 2\phi_{FB} - \frac{\gamma^2}{2} + \gamma\sqrt{-V_G + V_{FB} + \frac{\gamma^2}{4}}$$

A p-channel MOSFET is cut off for $V_G - V_S > V_T$. The strong inversion model predicts $I = 0$ for this mode.

As to the simplified models, the intermediate model predicts nonsaturation for

$$V_{GS} - V_T \leq (1 + \delta')V_{DS}$$

The associated channel current is described by

$$I = -K_P\frac{W}{L}\left[V_{GS} - V_T - \frac{1}{2}(1 + \delta')V_{DS}\right]V_{DS}$$

where

$$\delta' \equiv \frac{\gamma}{2(-V_S - 2\phi_{FB})^{1/2}}$$

The saturation condition of the same model is given by

$$(1 + \delta')V_{DS} < V_{GS} - V_T \leq 0$$

The current is described by

$$I = -\frac{1}{2}K_P\frac{W}{L}\frac{1}{1 + \delta'}(V_{GS} - V_T)^2$$

The simple model is represented by the following equations:

$$I = -K_P\frac{W}{L}\left(V_{GS} - V_T - \frac{1}{2}V_{DS}\right)V_{DS}$$

and

$$V_{GS} - V_T \leq V_{DS}$$

for nonsaturation, and

$$I = -\frac{1}{2}K_P\frac{W}{L}(V_{GS} - V_T)^2$$

and

$$V_{DS} < V_{GS} - V_T \leq 0$$

for saturation.

Finally, the p-channel version of the subthreshold model is given by

$$I = -K_P\frac{W}{L}\frac{(kTn_i/qN_D)^2}{\sqrt{1 - (4/\gamma^2)(V_G - V_{FB})} - 1}\exp\left\{\frac{q}{kT}\frac{\gamma^2}{4}\left[\sqrt{1 - \frac{4}{\gamma^2}(V_G - V_{FB})} - 1\right]^2\right\}$$
$$\left[\exp\left(\frac{q}{kT}V_S\right) - \exp\left(\frac{q}{kT}V_D\right)\right]$$

or, alternatively, by

$$I = -\frac{1}{2}K_P\frac{W}{L}\left(\frac{kT}{q}\frac{n_i}{N_D}\gamma\right)^2\frac{\exp[-(q/kT)(V_{GS} - V_{TX})]}{-V_{TX} + V_{FB}}\left[1 - \exp\left(\frac{q}{kT}V_{DS}\right)\right]$$

where

$$V_{TX} \equiv V_{FB} + \frac{\gamma^2}{2} - \gamma\sqrt{\frac{\gamma^2}{4} - V_{GS} - V_S + V_{FB}}$$

3.3 FUNDAMENTALS OF STRUCTURAL OPTIMIZATION

The basic MOSFET structure we have studied thus far cannot adequately satisfy the rather sophisticated set of device specifications demanded by present-day integrated-circuit applications. Several structural enhancements, increasing the degree of device designers' freedom, have been introduced in the past to overcome this problem. *Threshold adjustment* is by far the most common of these enhancements. In this section, we will examine the issues related to device optimization, show the reasons leading to a necessity of profiling the transversal doping concentration, and present an analysis of the threshold voltage in such profiled structures.

3.3.1. Design Constraints

The structural features of a MOSFET are determined by device designers in the transversal direction and by circuit designers in the lateral direction. Considering the simple strong inversion model equations presented in Section 3.2.4, we can immediately identify four structure-related parameters; namely, W, L, K_P, and V_T. Among these, only the first two are associated with the lateral geometry. These two have technologically defined lower limits, but otherwise are under full control of the circuit designer. Unless there is a compelling reason to do otherwise, the circuit designer minimizes one of these two and sets the other appropriately to obtain the desired transconductance in the smallest possible gate area $W \times L$ because the MOSFET current is proportional to W/L.

The two remaining parameters, K_P and V_T, being related to the transversal structural features, are under the control of device designers whose goals would usually be (1) to maximize K_P for achieving the highest possible transconductance, (2) to adjust the value of V_T to what is considered appropriate for the targeted technological family, and (3) to minimize the sensitivity of V_T to V_S for the reasons to be discussed soon. Let us now see how he or she could proceed toward achieving these goals.

The Transconductance Parameter K_P

The transconductance parameter K_P was defined in Section 3.2.3 for an n-channel MOSFET as

$$K_P \equiv \overline{\mu}_n C_{ox} = \overline{\mu}_n \frac{\epsilon_{ox}}{T_{ox}} \tag{3.91}$$

It has the dual definition

$$K_P \equiv \overline{\mu}_P C_{ox} = \overline{\mu}_P \frac{\epsilon_{ox}}{T_{ox}} \tag{3.92}$$

for a p-channel MOSFET. In these expressions, ϵ_{ox} and the effective mobilities, and $\overline{\mu}_n$, and $\overline{\mu}_p$, are physical constants. It is timely to note that these mobilities are not only one-third to one-half of the corresponding bulk mobilities because the Si-SiO$_2$ interface acts as an additional scattering medium but are also functions of the transversal field in the channel. This dependence is usually modeled by the empirical expression [11]

$$\overline{\mu} = \begin{cases} \overline{\mu}_o \left[\dfrac{U_c}{\mathscr{E}(0)} \right]^{U_e} & \text{for } \mathscr{E}(0) \geq U_c \\ \overline{\mu}_o & \text{for } \mathscr{E}(0) < U_c \end{cases} \tag{3.93}$$

where $\overline{\mu}_o$ is the low-transversal-field mobility, U_c ($\sim 10^4$ V/cm) is a *critical field* parameter, and U_e (~ 0.1) is an *exponential coefficient*. Considering that the surface potential is given by $\psi_s = V_S + 2\phi_{FB}$ for the source end of the channel and by $\psi_s = V_D + 2\phi_{FB}$ in nonsaturation or $\psi_S = V_{D(\text{sat})} + 2\phi_{FB}$ in saturation for the drain end, we obtain from (3.29) the following expressions for the transversal boundary field at the two ends of the channel:

$$\mathscr{E}_x(0,0) = \frac{C_{ox}}{\epsilon_s}(V_{GS} - V_{FB} - 2\phi_{FB})$$

$$\mathscr{E}_x(0,L) = \frac{C_{ox}}{\epsilon_s}(V_{GS} - V_{FB} - 2\phi_{FB} - V_{DS})$$

$$\mathscr{E}_x(0,L') = \frac{C_{ox}}{\epsilon_s}(V_{GS} - V_{FB} - 2\phi_{FB} - V_{DS(\text{sat})})$$

Quite obviously, a nonzero drain-source bias causes the field to vary in the lateral direction. To avoid the computational burden of this variation, the following simplified expression of an average field is usually used in (3.93):

$$\mathscr{E}(0) = \frac{C_{ox}}{\epsilon_s}|V_{GS} - V_T - U_t V_{DS}|$$

in nonsaturation or

$$\mathscr{E}(0) = \frac{C_{ox}}{\epsilon_s}|V_{GS} - V_T - U_t V_{DS(\text{sat})}|$$

in saturation where U_t (~ 0.5) is called the *transversal-field coefficient*.

Returning to (3.91) and (3.92), we identify the gate-oxide thickness T_{ox} as the only parameter that can be controlled by the device designer. Naturally, he or she would employ the minimum possible T_{ox} to maximize K_P. The lower limit of T_{ox} is determined by a number of considerations involving the stability and integrity of the oxide film, and can be as low as 10 nm in present-day technologies.

The Threshold Voltage V_T

The threshold voltage, originally defined by (3.56) in Section 3.2.3 for an *n*-channel MOSFET and then extended to a *p*-channel MOSFET in Section 3.2.6, can also be expressed as

$$V_T = V_{T0} + \gamma[(V_S + 2\phi_{FB})^{1/2} - (2\phi_{FB})^{1/2}] \tag{3.94}$$

$$V_{T0} \equiv V_{FB} + 2\phi_{FB} + \gamma(2\phi_{FB})^{1/2} \tag{3.95}$$

for an *n*-channel device, and as

$$V_T = V_{T0} - \gamma[(-V_S - 2\phi_{FB})^{1/2} - (-2\phi_{FB})^{1/2}] \tag{3.96}$$

$$V_{T0} \equiv V_{FB} + 2\phi_{FB} - \gamma(-2\phi_{FB})^{1/2} \tag{3.97}$$

for a *p*-channel device, where V_{T0} represents the threshold voltage for $V_S = 0$, and is called the *zero-bias threshold voltage*. The target value of V_{T0} is determined primarily by the technological family to which the device belongs. In the most popular of the present-day families, the CMOS, for example, V_{T0} is expected to be in the range of 0.5 to 1.0 V for the *n*-channel MOSFET and in the range of -0.5 to -1.0 V for the *p*-channel MOSFET. In the less popular NMOS-only family, the *n*-channel driver MOSFETs require the same range of V_{T0} values as their CMOS counterparts but the *n*-channel load MOSFETs must have a negative V_{T0} of several volts. Unfortunately, despite the rather rich appearance of (3.95) and (3.97) in terms of parameter variety, the device designer has only one primary structural parameter—the substrate doping concentration—at his or her disposal for implementing the target value of V_{T0}! To see why, first consider the flatband voltage defined by (3.30). Among its variables, Q_{ss} is usually negligibly small in present-day technologies; ϕ_{FG} is fixed at about -0.6 V by the requirement of a highly conductive and electrically stable gate electrode, which can be best fulfilled by employing a heavily *n*-type-doped polysilicon material; and ϕ_{FB} is only a very weak (logarithmic) function of the substrate doping concentration. Practically, therefore, V_{FB} is fixed at about -1 V for an *n*-channel MOSFET and -0.3 V for a *p*-channel MOSFET. With V_{FB} and ϕ_{FB} being virtually constant, γ is left in (3.95) and (3.97) as the only parameter the device designer can manipulate. Inspecting

(3.43), and assuming T_{ox} to be already minimized, we identify the substrate doping concentration as the sole primary structural parameter by which γ and hence V_{T0} can be adjusted.

Shown in Figure 3.13(a) is the typical zero-bias threshold voltage versus substrate doping concentration curves calculated for three different values of T_{ox}. With regard to the n-channel device, we observe that V_{T0} cannot be made more negative than about -0.5 V. Therefore, the load device of the NMOS family cannot be realized with the MOSFET structure we have studied thus far.

A second observation concerning the n-channel device is that to adjust V_{T0} to the 0.5- to 1.0-V range of the CMOS family, one must employ a doping concentration in excess of about 3×10^{16} cm^{-3} for a 40-nm-thick gate oxide. For a 10-nm oxide, the required concentration reaches mid-10^{17} cm^{-3} levels. Unfortunately, such high concentration levels have several detrimental effects on device performance. First of all, note from (3.94) that the sensitivity of V_T to V_S is directly proportional to γ, and the latter is an increasing function of the doping concentration as depicted in Figure 3.13(b). The doping concentration levels that are necessary for properly setting the value of V_{T0} result in γ values that are too high (in excess of 1 V$^{1/2}$) to satisfy the other design objective of a low threshold sensitivity to bias. A second hazard arises from the fact that, for a given set of bias conditions, the electric field in junction depletion regions and in the pinch-off region of saturation is higher for a heavier doping concentration. At those concentration levels required for properly setting V_{T0}, the field becomes sufficiently high to cause considerable impact ionization and to lower the threshold of avalanche breakdown. Finally, such high levels of doping concentration make junction capacitances large enough to impair severely the speed performance of the device.

What we also observe from Figure 3.13(a) is that the requirement of a p-channel V_{T0} within the range of -0.6 to -1.0 V can hardly be met. Even then, one has to employ a very light substrate doping concentration not exceeding the low end of the order of 10^{14} cm^{-3}. Such a light doping concentration is unacceptable for a VLSI device because it causes the depletion regions and the pinch-off region to be too wide to fit into a reasonably small area. Unless the designer allows for larger device dimensions and device-to-device spacing and, thus, impairs integration density, one or more of the following problems would arise: (1) The pinch-off region occupies a large portion of the source-drain spacing and expands rapidly with bias, causing excessive channel-length modulation, and thus, reducing the output resistance of the MOSFET. (2) The depletion regions of the two junctions merge beneath the surface, and create a buried conduction path, which cannot be controlled by the gate voltage. This so-called *punch-through* condition severely reduces the transconductance of the MOSFET. (3) The depletion regions of the neighboring devices merge and can create leakage paths.

It is obvious from the foregoing discussion that, merely by adjusting the gate-oxide thickness and the substrate doping concentration, a device designer cannot meet all of the design objectives. Clearly, he or she must be provided with more

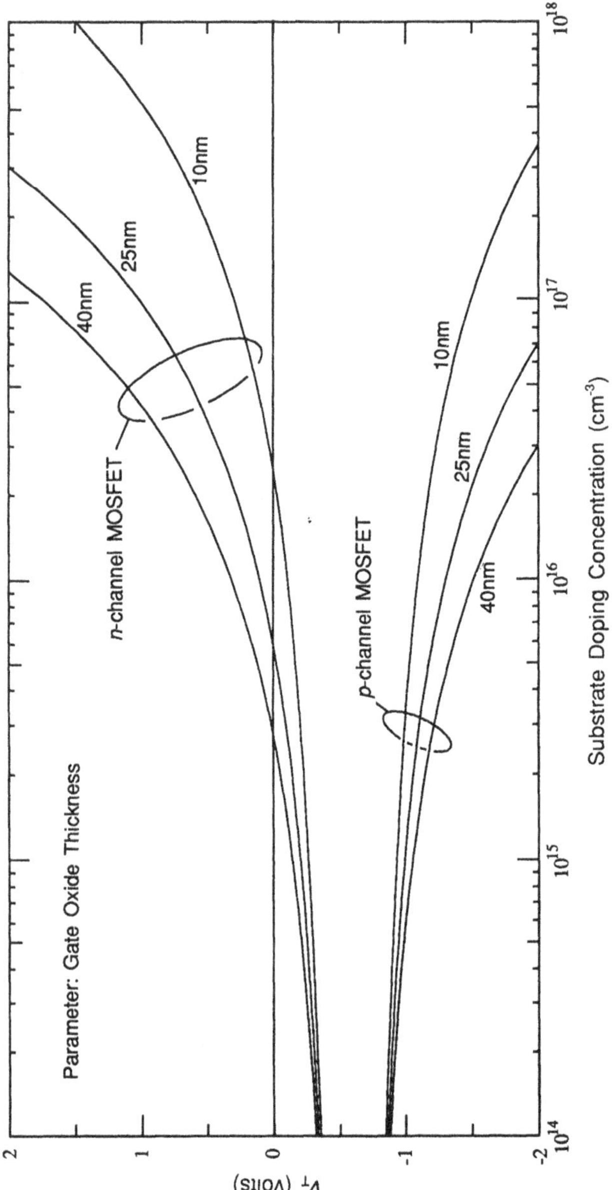

Parameter: Gate Oxide Thickness
n-channel MOSFET
p-channel MOSFET
10nm
25nm
40nm
10nm
25nm
40nm
V_T (volts)
2
1
0
-1
-2
10^{14}
10^{15}
10^{16}
10^{17}
10^{18}
Substrate Doping Concentration (cm^{-3})

(a)

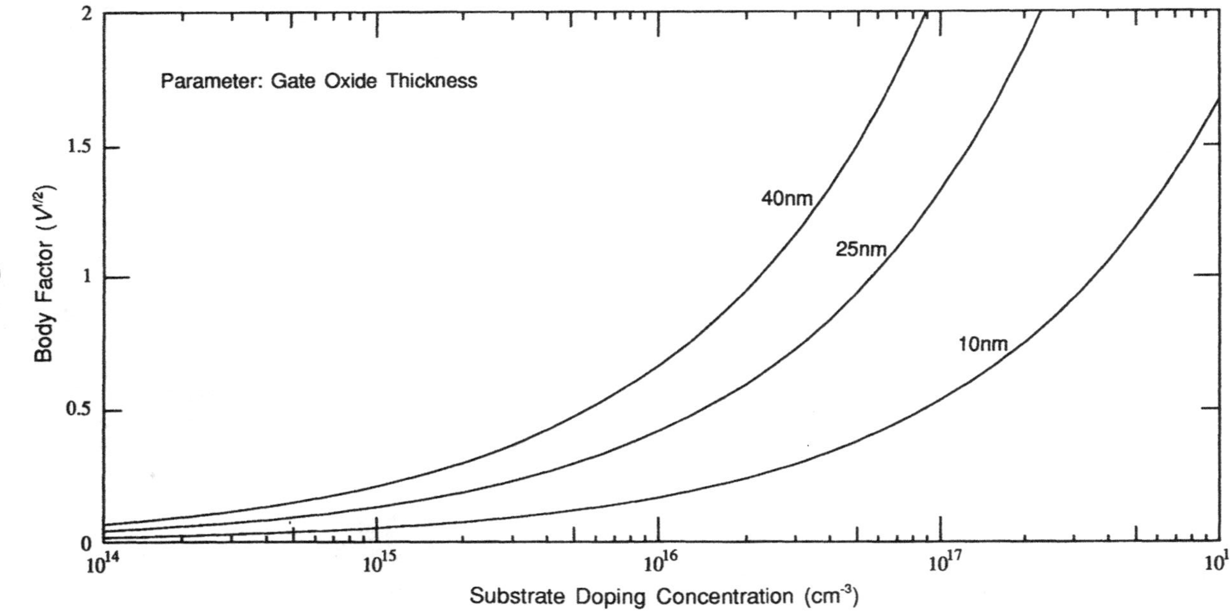

Figure 3.13 The effect of the substrate doping concentration on (a) zero-bias threshold voltage and (b) body factor.

degrees of freedom to accomplish this task, which is precisely why present-day MOSFETs have a carefully engineered nonuniform doping profile along the transversal direction. Profiling is done in two distinct ways, either by enhancing the surface doping concentration or by counter doping the surface. The resulting MOSFET structures are the subject of the following section.

3.3.2. Threshold Adjustment

MOSFETs With a Doping-Enhanced Channel

After having optimized the substrate doping concentration by an acceptable trade-off between the depletion-width-related effects favoring a high concentration and the threshold sensitivity, capacitance, and impact-ionization-related effects favoring a low concentration, one can independently shift V_{T0} in the positive direction in an *n*-channel device, or in the negative direction in a *p*-channel device, by increasing the doping concentration in the surface space-charge region only. A typical application of this technique is the threshold adjustment of the *n*-channel MOSFET of the *n*-well CMOS technology, in which the substrate doping concentration is in the mid-10^{15} cm^{-3} range. As indicated by Figure 3.13(a), the natural V_{T0} resulting from this concentration is close to 0V, whereas the target value is in the range of 0.5 to 1.0V, as mentioned previously. Therefore, a positive threshold shift of 0.5 to 1.0V is necessary.

For an analytical description of threshold shifting by doping enhancement in an *n*-channel device, consider the transversal *p*-type doping profile shown in Figure 3.14(a). It comprises a uniform N_A throughout the substrate and a surface enhancement N_T to a depth d_t [12]. Suppose that, under a threshold condition on the source side for which $V_G = V_S + V_T$, the depleted surface space-charge region has a sufficiently large thickness x_t to include the doping-enhanced region. The corresponding electric-field profile, as obtained by solving Poisson's equation, is shown in Figure 3.14(b), from which we can write the following equation for the boundary field:

$$\mathscr{E}(0) = \frac{q}{\varepsilon_s}(N_A x_t + N_T d_t) \tag{3.98}$$

Relying on the fact that the area bounded by the field profile defines the surface potential, whose threshold value is $V_S + 2\phi_{FB}$, we obtain the additional equation

$$x_t = \sqrt{\frac{2\epsilon_s}{qN_A}(V_S + 2\phi_{FB}) - \frac{N_T}{N_A} d_t^2} \tag{3.99}$$

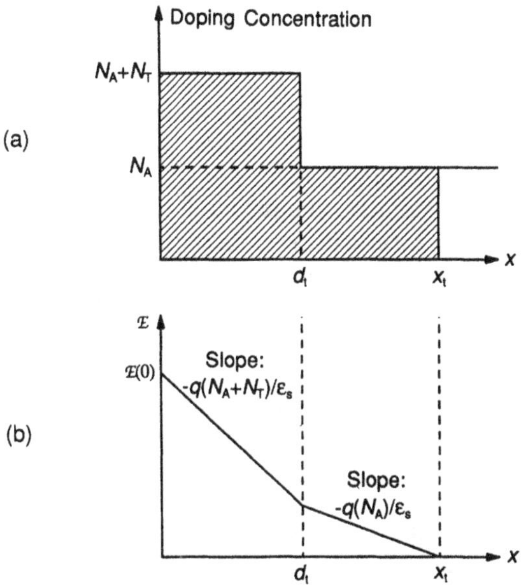

Figure 3.14 (a) The ideal transversal doping profile and (b) the electric-field profile in a doping-enhanced MOSFET.

whose substitution back into (3.98) yields the following expression of $\mathscr{E}(0)$ at the threshold:

$$\mathscr{E}(0) = \sqrt{\frac{2qN_A}{\epsilon_S}(V_S + 2\phi_{FB}) - \frac{N_T}{N_A}\left(\frac{qN_A d_t}{\epsilon_S}\right)^2} + \frac{qN_T d_t}{\epsilon_S}$$

Finally, using the latter in (3.29) together with $\psi_s = V_S + 2\phi_{FB}$ and $V_G = V_S + V_T$, we arrive at

$$V_T = V_{T0} + \gamma\left[\left(V_S + 2\phi_{FB} - \frac{qD_T d_t}{2\epsilon_s}\right)^{1/2} - \left(2\phi_{FB} - \frac{qD_T d_t}{2\epsilon_s}\right)^{1/2}\right] \quad (3.100)$$

where

$$V_{T0} \equiv V_{FB} + \frac{qD_T}{C_{ox}} + 2\phi_{FB} + \gamma\left(2\phi_{FB} - \frac{qD_T d_t}{2\epsilon_s}\right)^{1/2} \quad (3.101)$$

$$D_T \equiv N_T d_t$$

and γ is the body factor defined by (3.43). Note that D_T represents the adjustment doping concentration per unit area and is usually called the *dose*.

Compared with (3.95) of the uniform doping profile, (3.101) contains two additional terms. The first, qD_T/C_{ox}, is an additive term by which the desired threshold shift in the positive direction is implemented. The second, $qD_Td_t/2\epsilon_s$, is a subtractive term, which is counteractive not only in shifting the threshold in the right direction but also in reducing the sensitivity of V_T to V_S, as understood from (3.100). Fortunately, this term can be minimized by realizing the necessary D_T within the shortest possible range d_t. This is why the technological process of threshold-enhancement implantation is confined to a very narrow surface region.

Assuming that the counteractive term is thus minimized, we observe from (3.100) that the sensitivity of V_T to V_S is determined by the substrate doping concentration and not by the enhanced surface doping. Therefore, the device designer has an additional degree of freedom in setting the threshold voltage and body factor independently.

Finally, note that, other than adjusting the value of V_{T0}, the enhanced surface doping concentration has little, if any, effect on the characteristic equations describing the MOSFET current.

MOSFETs with a Counter-Doped Channel (Buried-Channel MOSFETs)

As we have just discussed, increasing the surface doping concentration results in a positive threshold shift in an *n*-channel device and a negative one in a *p*-channel device. Can we shift the threshold voltage in the other direction by reducing surface concentration, say, by counter doping the substrate? The answer is affirmative, but the counter-doping concentration required for any significant threshold shift is usually much larger than the substrate concentration. Inevitably, therefore, the type of conductivity is changed at the surface, resulting in a *pn* junction located just beneath the surface, as shown in Figure 3.15 for a *p*-channel device. The reversal of conductivity obviously results in a metallurgical channel connecting the source and drain junctions, which is why these types of devices are also called *buried-channel MOSFETs*. The operation of a buried-channel MOSFET is based on a majority-carrier drift along the channel. The upper edge of the depletion region of the channel-substrate junction defines the lower electrical boundary of the channel. Under surface-accumulating bias conditions, the channel of drifting majority carriers extends all the way up to the Si-SiO_2 interface, causing a large channel conductivity. If, however, a surface-depleting bias condition prevails, then the lower edge of this surface depletion region becomes the upper boundary of the channel, which, being thus constricted, exhibits reduced conductivity. This is how the channel current is controlled by the gate voltage. For the operational details and modeling of buried-channel MOSFETs the reader is referred to Van

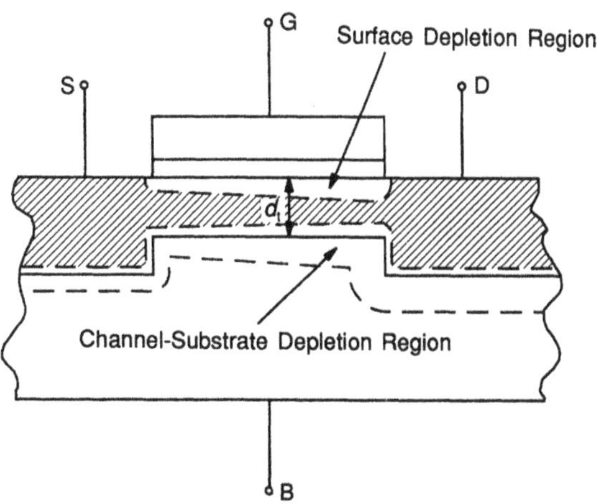

Figure 3.15 The channel region in a counter-doped (buried-channel) MOSFET.

Der Tol and Chamberlain [13]. Here, we will discuss only the threshold voltage of these devices because, except for the physically different makeup of this parameter, the modeling equations developed for uniformly doped MOSFETs, particularly the simplified ones presented in Section 3.2.4, are applicable to buried-channel devices as well.

The threshold voltage of a buried-channel MOSFET is defined as the difference $V_G - V_S$ that causes the surface depletion region to meet the depletion region of the channel-substrate junction at the source side. If d_t, x_{ts}, and x_{tj} denote the thicknesses of the counter-doped surface region, the surface depletion region, and the junction depletion region inside the counter-doped region, respectively, then the threshold condition implies

$$d_t - x_{ts} - x_{tj} = 0 \tag{3.102}$$

Assuming a p-channel device, we can adopt (3.46) to formulate x_{ts} because that equation describes the thickness of a surface depletion region located in any uniformly doped p-type material, which is what the counter-doped region of a p-channel buried-channel MOSFET indeed is. However, the V_G of that equation, being defined as the gate-substrate voltage, must now be replaced by $V_G - V_S$ for calculating the depletion thickness on the source side. In turn, $V_G - V_S$ equals V_T under the threshold condition. We, therefore, obtain

$$x_{ts} = -\frac{\epsilon_s}{C_{ox}} + \sqrt{\left(\frac{\epsilon_s}{C_{ox}}\right)^2 + \frac{2\epsilon_s}{qN_A}(V_T - V_{FB})} \tag{3.103}$$

As to the junction depletion thickness inside the p-type counter-doped region, x_{tj}, we can adopt the solution of Problem 1.2, part (d), where x_2 and V are analogous to x_{tj} and V_S. Assuming that the doping concentration N_A of the counter-doped region is much larger than the N_D of the n-type substrate, the result will be

$$x_{tj} = \sqrt{\frac{2\epsilon_s}{q}\frac{N_D}{N_A^2}(V_b - V_S)} \tag{3.104}$$

where the junction built-in voltage V_b is defined by

$$V_b \equiv \frac{kT}{q}\ln\frac{N_A N_D}{n_i^2}$$

Substituting (3.103) and (3.104) into (3.102) and solving for V_T, we obtain

$$\begin{aligned} V_T = V_{FB} &+ \frac{qD_T}{C_{ox}}\left(1 + \frac{d_t C_{ox}}{2\epsilon_s}\right) + \frac{N_D}{N_A}(V_b - V_S) \\ &- \left(1 + \frac{d_t C_{ox}}{\epsilon_s}\right)\gamma\sqrt{V_b - V_S} \end{aligned} \tag{3.105}$$

where $D_T \equiv N_A d_t$ is the dose of the counter doping, and

$$\gamma \equiv \frac{1}{C_{ox}}\sqrt{2q\varepsilon_s N_D}$$

is the body factor of the substrate. Neglecting the third term on the right-hand side of (3.105) because $N_A \gg N_D$, we finally arrive at

$$V_T = V_{T0} - \left(1 + \frac{d_t C_{ox}}{\epsilon_s}\right)\gamma(\sqrt{V_b - V_S} - \sqrt{V_b}) \tag{3.106}$$

where the zero-bias threshold voltage is defined by

$$V_{T0} \equiv V_{FB} + \frac{qD_T}{C_{ox}}\left(1 + \frac{d_t C_{ox}}{2\epsilon_s}\right) - \left(1 + \frac{d_t C_{ox}}{\epsilon_s}\right)\gamma\sqrt{V_b} \tag{3.107}$$

The second term on the right-hand side of (3.107) represents the positive shift in the threshold voltage. Also, note from (3.106) that, due to the finite thickness d_t of the counter-doped region, the effective body factor of a buried-channel MOSFET is somewhat larger than the body factor determined by the substrate alone.

3.4. SECONDARY EFFECTS

The MOSFET model developed in previous sections is based on a number of simplifying assumptions, some of which become increasingly unjustifiable as the device dimensions are reduced or bias levels are elevated. As a result, certain physical phenomena observed under these conditions in real devices are not adequately described by the modeling equations. In this section, we will identify these phenomena, discuss their effect on device operation, and show the ways to incorporate them in MOSFET models.

3.4.1. Velocity Saturation

The MOSFET model development presented in Section 3.2.3 is based on an integration of the basic differential equation (3.19) assuming that the effective mobility $\bar{\mu}$ is independent of the lateral field $\mathscr{E}_y$. This assumption is justifiable only in long-channel devices operating under relatively low levels of drain-source bias, in which the lateral field is weak. As L is decreased or V_{DS} is increased, $\mathscr{E}_y$ becomes strong enough to reduce $\bar{\mu}$, which makes I smaller than that predicted by the constant-mobility-based models [14].

The lateral-field dependence of the effective mobility can be accounted for in the modeling equations by adopting (2.112) for $\bar{\mu}$, that is,

$$\bar{\mu} = \frac{v_s}{\mathscr{E}_o + (dV/dy)}$$

where v_s is the magnitude of the saturation velocity, $\mathscr{E}_o$ is the critical field parameter, and dV/dy represents $-\mathscr{E}_y$. Substituting this equation into (3.19) for $\bar{\mu}_n$, rearranging, integrating along the channel as done in (3.51), and considering that $v_s/\mathscr{E}_0$ equals the low-field mobility $\bar{\mu}_{n0}$, we obtain

$$I \int_0^L dy = W\bar{\mu}_{n0} \int_{V_S}^{V_D} Q_n \, dV - \frac{I}{\mathscr{E}_o} \int_{V_S}^{V_D} dV$$

for a nonsaturated MOSFET. Replacing the first and third integrals with L and V_{DS}, respectively, rearranging into

$$I\left(1 + \frac{V_{DS}}{\mathscr{E}_o L}\right) = \frac{W}{L} \bar{\mu}_{n0} \int_{V_S}^{V_D} Q_n \, dV$$

and noting that the right-hand side is the low-field current given by any of the models developed in Sections 3.2.3 through 3.2.5, we finally arrive at

$$I = \frac{I_0}{1 + (V_{DS}/\mathscr{E}_o L)} \tag{3.108}$$

where I_0 is the low-field current. This equation clearly shows how the current can be overestimated by the low-field models in a short-channel MOSFET. Assuming $\mathscr{E}_o = 6 \times 10^3$ V/cm, a 1-μm-long device would conduct only one-half the estimated current for a drain bias of just $V_{DS} = 0.6$V! What is not all that clear from (3.108) is how the lateral degradation of mobility modifies the characteristic equations in *saturation*. To see how, let us use the simplest saturation equation (3.79) for I_0 in (3.108), and solve the equation $dI/dV_{DS} = 0$ for V_{DS}. This gives us the drain-source bias for which I saturates, that is, $V_{DS(\text{sat})}$, as follows:

$$V_{DS(\text{sat})} = \mathscr{E}_o L\left(\sqrt{1 + 2\frac{V_{GS} - V_T}{\mathscr{E}_o L}} - 1\right) \tag{3.109}$$

which indicates a smaller $V_{DS(\text{sat})}$ than $V_{DS(\text{sat})} = V_{GS} - V_T$ of the basic model in which the lateral degradation of $\bar{\mu}$ is ignored. At this drain bias, the carriers attain saturation velocity at the drain side of the channel, which is why the effect of lateral degradation of mobility is more commonly known as the *velocity saturation effect*.

Substituting (3.79) for I_0 into (3.108), and then using (3.109) for V_{DS}, we can obtain the characteristic equation describing I in saturation. Again a current smaller than the prediction of the basic model is indicated. For a high gate bias satisfying $V_{GS} - V_T \gg \mathscr{E}_o L/2$ for which (3.109) reduces to $V_{DS(\text{sat})} \cong \sqrt{2\mathscr{E}_o L(V_{GS} - V_T)}$, the saturation current becomes

$$I \cong W v_s C_{ox}(V_{GS} - V_T) \tag{3.110}$$

Two interesting conclusions can be drawn from this equation for the consequences of the velocity saturation effect under high gate bias: (1) The current becomes linearly dependent on the gate voltage in contradistinction to the quadratic dependence observed for a weak field and (2) the channel length L ceases to be a parameter of the current!

3.4.2. Channel-Length Modulation

As described in Section 3.2.3, the channel-length modulation effect is due to an expansion of the pinch-off region of saturation with drain bias. Since it shortens the effective channel length, the channel current becomes an increasing function of the drain voltage in saturation. A finite output resistance is therefore implied.

To characterize the effect of channel-length modulation on the channel current, one has to find a relationship between the length ΔL of the pinch-off region and the drain voltage V_D. Unfortunately, an accurate relationship can only be obtained from a two-dimensional, hence numerical, solution of Poisson's equation because the pinch-off region is under the influence of a field distribution of comparable transversal and lateral components. Analytical formulations based on one-dimensional or quasi-two-dimensional solutions can provide only crude estimates for ΔL, but are nevertheless useful in drawing a qualitative picture of the effect. The most popular of these is based on a one-dimensional solution of Poisson's equation in the lateral direction under the conditions depicted in Figure 3.16. Note the following assumptions:

1. The pinch-off region extending between $y = L'$ and $y = L$ is depleted.
2. The width of the depletion region inside the drain is negligibly small because the drain is very heavily doped.
3. The boundary values of the lateral field are given by $\mathscr{E}_y(L') = -\mathscr{E}_T$ and $\mathscr{E}_y(L^+) = 0$, where L^+, being the location of the depletion edge inside the heavily doped drain, is practically equal to L.

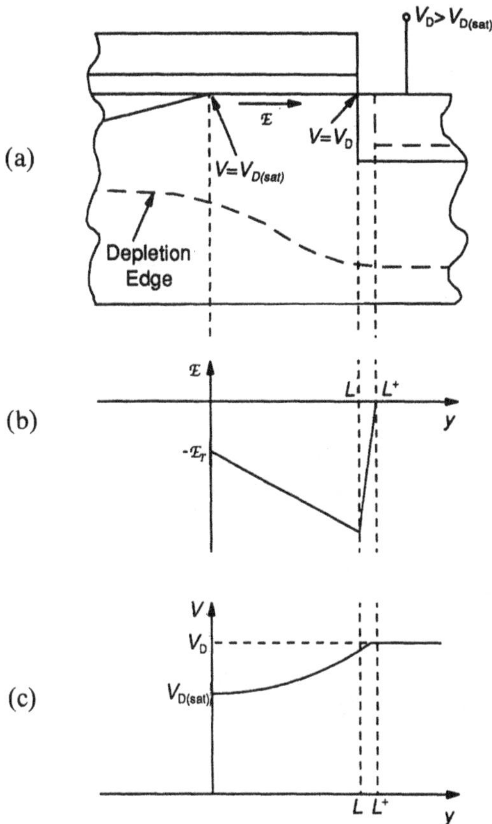

Figure 3.16 A one-dimensional model for the channel-length modulation effect: (a) The geometry of the pinch-off region, (b) electric-field profile, and (c) channel potential profile.

4. The potential V varies between $V_{D(\text{sat})}$ at $y = L'$ and V_D at $y = L^+$. The solution for ΔL is as follows:

$$\Delta L = \sqrt{\frac{2\epsilon_s}{qN_A}} \left(\sqrt{V_{DS} - V_{DS(\text{sat})} + \phi_T} - \sqrt{\phi_T}\right) \qquad (3.111)$$

where ϕ_T is defined by

$$\phi_T \equiv \frac{\epsilon_s \mathscr{E}_T^2}{2qN_A}$$

and $\mathscr{E}_T$ is treated as an empirical parameter [15].

Substituting (3.111) for ΔL in $L' = L - \Delta L$, and using the latter in the appropriate saturation equation, (3.66), (3.67), (3.78) or (3.81), we can incorporate the channel-length modulation effect into any of the strong inversion models presented in Sections 3.2.3 and 3.2.4.

Another popular form in which the channel-length modulation effect is represented in modeling equations is

$$I = \frac{I_0}{1 - \lambda V_{DS}}$$

where I_0 is the saturated channel current as given by any of the above-cited basic model equations for $L' = L$, and λ is the so-called *channel-length modulation parameter* [16]. Obviously, the actual current is related to I_0 by

$$I = \frac{L}{L - \Delta L} I_0$$

Equating the right-hand sides of these two equations, and substituting (3.111) for ΔL, we can express λ in terms of V_{DS} and other parameters and constants appearing in (3.111). The result is

$$\lambda = \frac{1}{L} \sqrt{\frac{2\epsilon_s}{qN_A}} \frac{\left(\sqrt{V_{DS} - V_{DS(\text{sat})} + \phi_T} - \sqrt{\phi_T}\right)}{V_{DS}}$$

which clearly shows that the severity of the channel-length modulation effect increases with a decreasing channel length L.

3.4.3 Punch-Through

In short-channel MOSFETs fabricated on lightly doped substrates, the drain depletion region expands rapidly with V_D in the substrate and eventually reaches the edge of the source depletion region. This physical situation, called *punch-through*, is undesirable because it reduces the potential barrier across the source junction below the surface, and thus causes an electron injection from source to drain. The resulting current is regarded as a leakage because it can hardly be controlled by the gate voltage [17].

A typical pre-punch-through condition of the source and drain depletion regions is schematically illustrated in Figure 3.17(a). For the sake of analytical simplicity we assume an n-channel MOSFET operating with $V_S = 0$. Notice that beneath the surface space-charge layer, a p-type neutral bulk separates the depletion regions of the two junctions. The electric-field profile and energy-band diagram associated with the lateral path S-D will therefore resemble those shown in Figure 3.17(b, c), respectively.

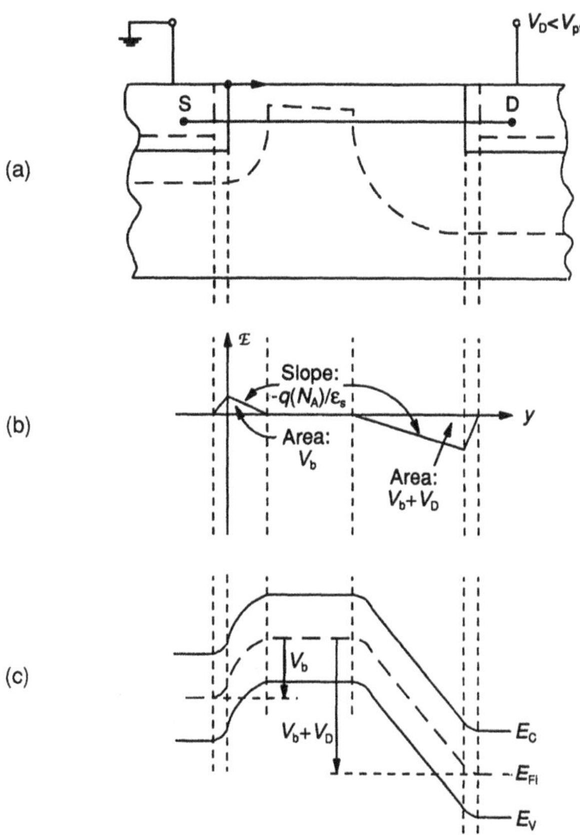

Figure 3.17 A typical pre-punch-through condition in the channel: (a) spreading of the depletion regions, (b) electric-field profile, and (c) energy-band diagram.

Now suppose that V_D is increased. The triangular area bounded by the field profile of the drain depletion region will expand and thus push the depletion edge toward that of the source depletion region. Eventually, the two edges meet as shown in Figure 3.18(a). This situation is called the onset of punch-through, whereas the value of V_D that causes this situation is known as the *punch-through voltage*, V_{pt}. The corresponding field profile and energy-band diagram are given in Figure 3.18(b, c), respectively. Using the simple geometrical features of the former, we can easily obtain the following expression for V_{pt}:

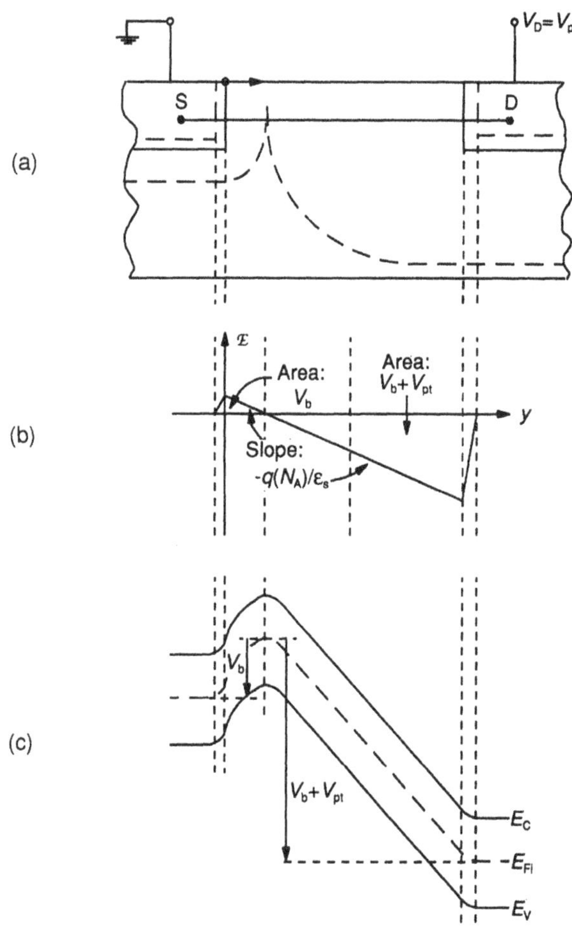

Figure 3.18 Onset of punch-through: (a) spreading of the depletion regions, (b) electric-field profile, and (c) energy-band diagram.

$$V_{pt} = \frac{1}{2}\frac{kT}{q}\left(\frac{L}{L_D}\right)^2 - \frac{L}{L_D}\sqrt{2\frac{kT}{q}V_b} \tag{3.112}$$

where

$$L_D \equiv \sqrt{\frac{kT\epsilon_s}{q^2N_A}}$$

is the substrate Debye length, and V_b is the junction built-in potential.

At the onset of punch-through the source junction still remains in a thermal equilibrium state. Its depletion region, extending to the point $y = y_m$ where the potential minimum is located, does not yet support a net current. Stated alternatively, the opposing diffusion and drift currents inside this region are still balanced. As V_D is increased above V_{pt}, however, the balance is upset in favor of the diffusion current because the location of the potential minimum moves toward the source, reducing the width of the source depletion region and also the field strength inside this region, as depicted in Figure 3.19. Also note that the potential barrier becomes smaller than V_b, which is usually identified as the *drain-induced barrier lowering* (DIBL) effect. The electrons, now diffusing across the source depletion region, are injected into the drain depletion region at $y = y_m$, beyond which a strong negative field sweeps them into the drain. This is how the punch-through current is generated.

As a first-order approximation in modeling the punch-through current J_{pt}, we can ignore the transversal field and regard the punch-through current as a purely lateral current, as shown in Figure 3.19(a). Using the general current density equation, we can express J_{pt} as

$$J_{pt} = q\mu_n n\mathscr{E} + qD_n\frac{dn}{dy}$$

Rearranging this equation, and integrating along y between $y = 0$ and $y = L$, we obtain

$$J_{pt}\int_0^L \frac{dy}{n} = q\mu_n\int_0^L \mathscr{E}\,dy + q\mu_n\frac{kT}{q}\int_{n(0)}^{n(L)}\frac{dn}{n}$$

The first integral on the right-hand side of this equation equals $-V_D$, whereas the second integral vanishes because $n(L) = n(0)$. This brings us to

$$J_{pt} = -\frac{q\mu_n V_D}{\displaystyle\int_0^L \frac{dy}{n}} \tag{3.113}$$

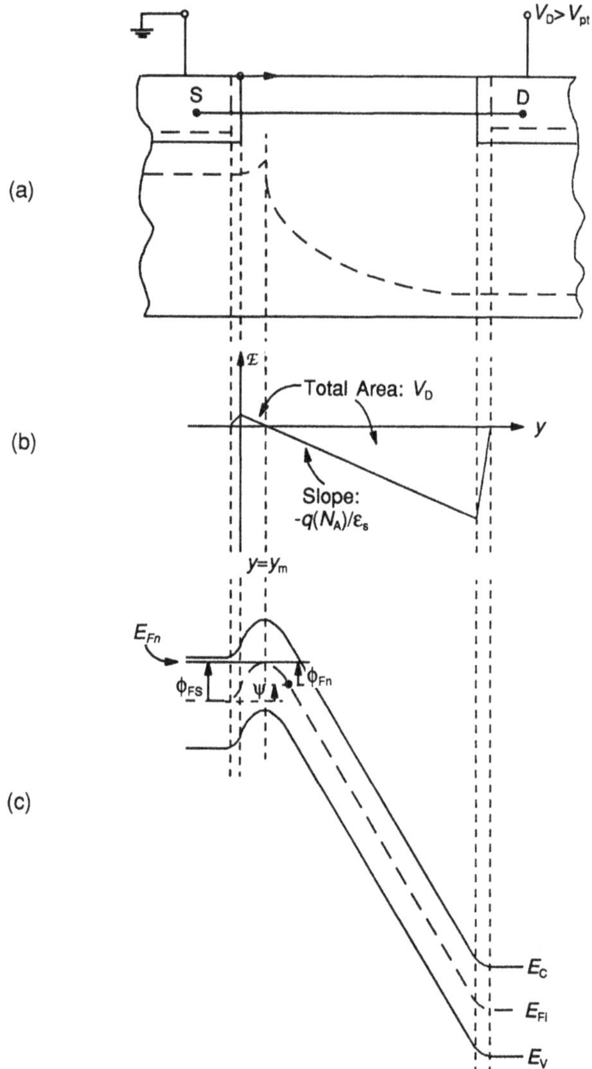

Figure 3.19 A typical punch-through condition: (a) spreading of the depletion regions, (b) electric-field profile, and (c) energy-band diagram.

For calculating the denominator integral, first we write

$$n = n_i \exp\left(-\frac{q}{kT}\phi_{Fn}\right)$$

where ϕ_{Fn} is the electron quasi-Fermi potential. Then, we make two observations: First, the value of the integral is determined mostly by the low electron concentration localized to a narrow region around $y = y_m$ and, second, this location is close to the source. Based on these observations, we can assume, as a first-order approximation, a position-independent electron quasi-Fermi level throughout this narrow region, as depicted in Figure 3.19(c). This enables us to write the loop equation $\phi_{Fn} = \phi_{FS} - \psi$, where ϕ_{FS} is the equilibrium Fermi potential of the source, and ψ is the electrostatic potential, for which E_{Fi} in the source is selected as the reference level. Substituting this equation into the previous one for ϕ_{Fn}, we obtain

$$n = n_i \exp\left(-\frac{q}{kT}\phi_{FS}\right)\exp\left(\frac{q}{kT}\psi\right) = N_S \exp\left(\frac{q}{kT}\psi\right) \tag{3.114}$$

where N_S is the source doping concentration. The electrostatic potential profile can be obtained by spatially integrating the field profile of Figure 3.19(b). The result is given by the following two equations:

$$\psi = \frac{kT}{q}\frac{1}{L_D^2}\left(\frac{1}{2}y^2 - y_m y\right) \tag{3.115}$$

$$y_m = \frac{1}{2}L - \frac{L_D^2}{L}\frac{q}{kT}V_D \tag{3.116}$$

Substituting (3.115) into (3.114) and using the letter in the denominator integral of (3.113), we arrive at

$$\int_0^L \frac{dy}{n} = \frac{1}{N_S}\int_0^L \exp\left[-\frac{1}{L_D^2}\left(\frac{1}{2}y^2 - y_m y\right)\right] dy$$

Applying the transformation $z = (y - y_m)/L_D$, we calculate this integral as

$$\int_0^L \frac{dy}{n} = \frac{L_D}{N_S}\exp\left[\left(\frac{y_m}{L_D}\right)^2\right]\int_{-y_m/L_D}^{(L-y_m)/L_D} \exp(-z^2)\, dz = \sqrt{\pi}\frac{L_D}{N_S}\exp\left[\left(\frac{y_m}{L_D}\right)^2\right]$$

Using (3.116) for y_m in this expression, and substituting the latter back into (3.113), we finally obtain

$$J_{pt} = -\frac{q\mu_n N_S}{\sqrt{\pi}L_D}V_D \exp\left[-\left(\frac{L}{2L_D} - \frac{L_D}{L}\frac{q}{kT}V_D\right)^2\right] \tag{3.117}$$

which indicates a current flowing from drain to source and exponentially increasing with the drain bias. The relative bias sensitivity of this current is given by

$$\frac{d|J_{pt}|}{|J_{pt}|dV_D} = \frac{1}{V_D} + 2\frac{L_D}{L}\frac{q}{kT} \tag{3.118}$$

To suppress the punch-through effect, the device designer has to ensure a large V_{pt} and a small bias sensitivity. It is obvious from (3.111) and (3.118) that both objectives can be reached only by decreasing the ratio L_D/L, which calls for an increasingly large substrate doping concentration as the channel length is reduced. In present-day MOSFETs, the concentration required for this purpose is much larger than what can be tolerated beneath the source and drain junctions because the resulting junction capacitance would be excessively large. To avoid this conflict, the substrate concentration is enhanced by what is called *punch-through stopper implantation* to a depth not exceeding the bottom of the junctions. This enhancement effectively blocks out the punch-through path but still leaves the bottom areas of the junction exposed to a more lightly doped substrate, which yields a smaller junction capacitance.

The foregoing analysis of the punch-through effect is based on the assumption of $V_S = 0$ and a negligible transversal field. With regard to the effect of $V_S \neq 0$, it is worth knowing that a reverse bias can strongly suppress the punch-through current. On the other hand, the interaction between the transversal field and the punch-through effect is relatively weak because punch-through can occur at depths beyond the reach of the surface space-charge region.

3.4.4 Short-Channel and Narrow-Channel Effects

The surface space-charge analysis presented in Section 3.2.2 was based on the assumption of one-dimensional electric-field distribution along the transversal direction. The threshold voltage of (3.42), having been derived on this basis, is therefore valid only in those channel areas where no lateral field gradient exists. In long and wide MOSFETs most of the channel area indeed operates with a negligible lateral-field gradient but this situation changes rapidly as the lateral device dimensions are reduced below certain limits.

In a narrow-channel MOSFET, for example, the infringement of the surface space-charge region beyond the structurally defined width may become comparable to W, as shown in Figure 3.20(a). This results in an outward bending of field lines, hence, a significant lateral field distribution at the expense of the transversal field. To increase the transversal field to the level of threshold, one therefore needs to apply a greater gate voltage. In other words, the threshold voltage of a narrow-channel device is larger than what is predicted by the basic model. This is generally referred to as the *narrow-channel effect*.

In a short-channel MOSFET, on the other hand, the surface space-charge region can be influenced by a lateral field distribution due to the proximity of the

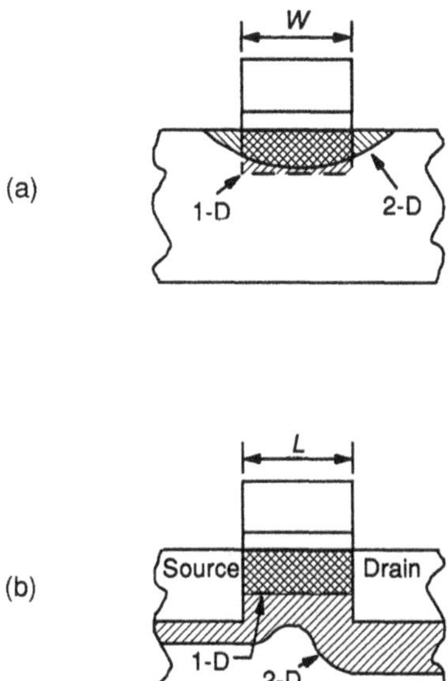

Figure 3.20 The actual two-dimensional (2-D) formation of the surface space-charge region in comparison with the ideal one-dimensional (1-D) formation: (a) in a narrow-width MOSFET and (b) in a short-channel MOSFET.

source and drain junctions, as illustrated in Figure 3.20(b). The lateral fields, being directed inward from both ends of the channel, contribute to the formation of a surface depletion region, so that a lesser transversal field, hence a smaller gate voltage, is needed for bringing the device to the onset of inversion. For this reason, the threshold voltage is a decreasing function of L in short-channel MOSFETs. This is generally known as the *short-channel effect.*

An accurate analysis of the narrow and short channel effects calls for a solution of the two-dimensional form of Poisson's equation with appropriate boundary values, which is possible only by numerical means. For this reason, the existing analytical models have been based on quasi-two-dimensional approaches, the most popular of which is the concept of *charge sharing*. In this approach, the gate, gate oxide, and the part of the silicon surface charged by the gate are assumed to constitute a volume in which the sum of all charges is zero according to Gauss's theorem, that is,

$$Q_{Gt} + Q_{SSt} + Q_{St} = 0 \tag{3.119}$$

where Q_{Gt}, Q_{SSt}, and Q_{St} are the gate charge, oxide fixed charge, and the gate-induced surface charge, respectively. It is important not to confuse these quantities with charge density. They simply represent charge as measured in units of coulomb. The derivation proceeds as follows.

First, consider the gate charge, which resides at the gate-SiO_2 interface, to be a sheet charge. Relying on the continuity of the displacement vector at this interface, and assuming a negligible field in the gate, we can express the areal density of this charge as $Q_G = \epsilon_{ox}\mathscr{E}_{ox}$, where $\mathscr{E}_{ox}$ is the oxide field. With the aid of (3.26) and (3.29), this equation can be rewritten

$$Q_G = C_{ox}(V_G - V_{FB} - \psi_s) - Q_{SS}$$

Now, multiplying both sides of this equation by the gate area $W \times L$, and replacing V_G and ψ_s by $V_S + V_T$ and $V_S + 2\phi_{FB}$, respectively, as required by the threshold condition, we obtain

$$Q_{Gt} = WL[C_{ox}(V_T - V_{FB} - 2\phi_{FB}) - Q_{SS}] \tag{3.120}$$

for the total gate charge at threshold.

Next, we consider the gate-induced surface charge in silicon, which is due only to dopant ions under depletion and threshold conditions. Assuming a depleted silicon volume of U, we therefore can write $Q_{st} = -qN_AU$. Let us now partition U as $U = WLx_t + \Delta U$, where the first term on the right-hand side represents the depleted volume as predicted by the basic one-dimensional analysis, whereas the second term represents the additional depletion volume resulting from the lateral field distribution. Based on the qualitative description given at the beginning of this section, we expect a positive ΔU in the case of a narrow-channel effect, and a negative one in the case of a short-channel effect. Adopting (3.47) for x_t, we therefore can express Q_{st} at threshold as follows:

$$Q_{st} = -qN_A\left[WL\sqrt{\frac{2\epsilon_s}{qN_A}(V_S + 2\phi_F)} + \Delta U\right]$$

Substituting this equation together with $Q_{SSt} = WLQ_{SS}$ and (3.120) into (3.119) and rearranging, we obtain the following expression for the threshold voltage:

$$V_T = V_{FB} + 2\phi_{FB} + \gamma\left(1 + \frac{\Delta U}{U_{1\text{-D}}}\right)\sqrt{V_S + 2\phi_{FB}} \tag{3.121}$$

where $U_{1\text{-D}} \equiv WLx_t$ is the depleted volume as predicted by the one-dimensional model. According to (3.121), the two-dimensional effects are represented by a modified body factor

$$\gamma^* \equiv \left(1 + \frac{\Delta U}{U_{1\text{-D}}}\right)\gamma \tag{3.122}$$

which is larger than γ for $\Delta U > 0$ (narrow-channel effect) and smaller than γ for $\Delta U < 0$ (short-channel effect). Therefore, the predicted threshold voltage is an increasing function of W and a decreasing function of L.

Equation (3.122) is the basis of all charge-sharing models. These models vary in the way they relate the ratio $\Delta U/U_{1\text{-D}}$ to structural parameters and bias voltages. We now consider two exemplar models, one of which was developed by Yau [18] for the short-channel effect and the other by Akers [19] for the narrow-channel effect.

The geometrical formulation of ΔU in Yau's model is illustrated in Figure 3.21(a), where r_j denotes the radius of source and drain junction curvature, and x_{tS} and x_{tD} are the depletion widths of these two junctions. The sum of the volumes associated with the two shaded triangular regions are assumed to constitute a negative ΔU. Therefore,

$$\Delta U = -\frac{W}{2}x_t(\Delta L_S + \Delta L_D)$$

and

$$\frac{\Delta U}{U_{1\text{-D}}} = -\frac{\Delta L_S + \Delta L_D}{2L}$$

(a)

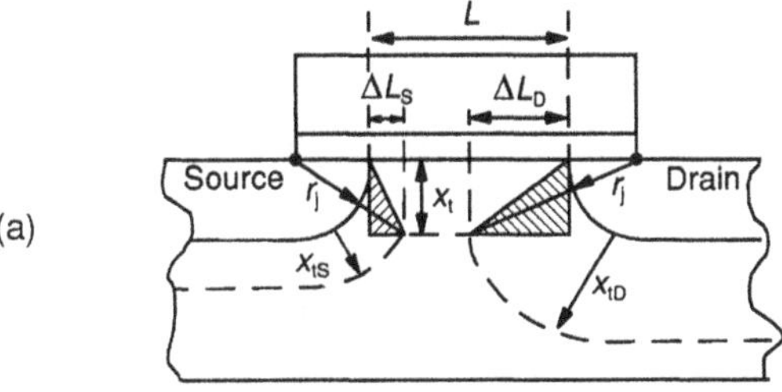

(b)

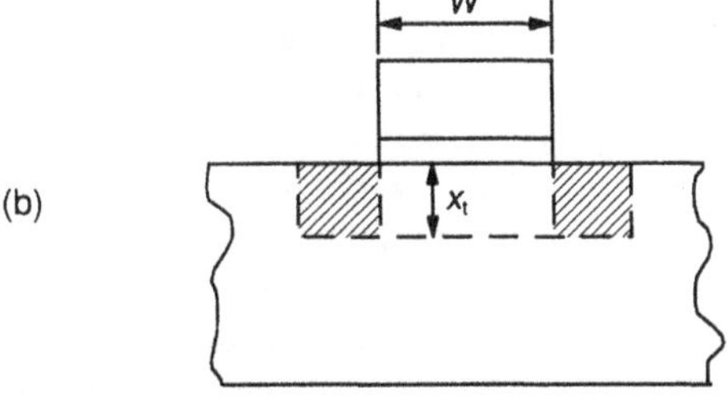

Figure 3.21 Geometric basis of (a) Yau's model [18] for the short-channel effect and (b) Akers' model [19] for the narrow-channel effect.

where

$$\Delta L_S = \sqrt{(r_j + x_{tS})^2 - x_t^2} - r_j$$
$$\Delta L_D = \sqrt{(r_j + x_{tD})^2 - x_t^2} - r_j$$

and

$$x_{tS} = \sqrt{\frac{2\epsilon_s}{qN_A}(V_S + V_b)}$$
$$x_{tD} = \sqrt{\frac{2\epsilon_s}{qN_A}(V_D + V_b)}$$
$$x_t = \sqrt{\frac{2\epsilon_s}{qN_A}(V_S + 2\phi_{FB})}$$

Note that the model predicts a threshold voltage reduction that is not only inversely proportional to L but is also an increasing function of junction bias voltages and the radius of curvature.

The model offered by Akers for the narrow-channel effect is based on a square spreading of the surface depletion region on both sides of the gate-defined width, as shown in Figure 3.21(b). The positive ΔU associated with these regions is given by $\Delta U = 2x_t^2 L$, which yields

$$\frac{\Delta U}{U_{\text{1-D}}} = 2\frac{x_t}{W}$$

indicating a threshold enhancement that is inversely proportional to the width.

3.4.5 Impact Ionization and Avalanche Breakdown

In short-channel MOSFETs with a reduced source-drain spacing and a relatively large channel doping concentration, the electric field inside the pinch-off region can easily reach sufficiently large levels to support impact ionization even under moderate drain-bias conditions. In an n-channel device, the electrons generated by impact ionization join those arriving from the source and move into the drain; the drain terminal current is thus increased, and the saturation property of the I_D-V_{DS} characteristics is impaired. The holes generated in the process of impact ionization flow to the substrate and thus give rise to a positive potential difference along the bulk in the transversal direction. This potential difference, by forward biasing the source-substrate junction, can create a positive feedback effect in two distinct ways: by causing further electron injection to the drain and by reducing the threshold voltage through body effect, which enhances the channel current. The positive feedback, thus established, leads to avalanching and, eventually, to a premature breakdown of the drain junction [20].

We will now present an analysis of impact ionization in short-channel MOSFETs assuming threshold lowering as the sole regenerative mechanism. Referring to Figure 3.22, we can express the drain and substrate currents as

$$\begin{aligned} I_D &= MI \\ I_{sub} &= (M - 1)I \end{aligned} \tag{3.123}$$

where M is the multiplication factor representing the effect of impact ionization in the pinch-off region, where the field is highest, and I is the channel current, which is given by (3.110) under extreme velocity saturation conditions. What is unique in the present case, however, is that the device operates with a negative source-substrate bias because, while the source is grounded, the substrate operates at the positive potential V_B due to the substrate current, that is,

$$V_B = R_B I_{sub} = (M - 1)R_B I \tag{3.124}$$

As a result, the threshold voltage is lowered in accordance with

$$V_T = V_{FB} + 2\phi_{FB} - K(V - 2\phi_{FB})$$

which is a better modeling equation than (3.56) for the present case of a forward-biased source. From this equation and (3.110), we can write

$$I = Wv_s C_{ox}[V_{GS} - V_{FB} - 2\phi_{FB}(1 + K) + KV_B]$$

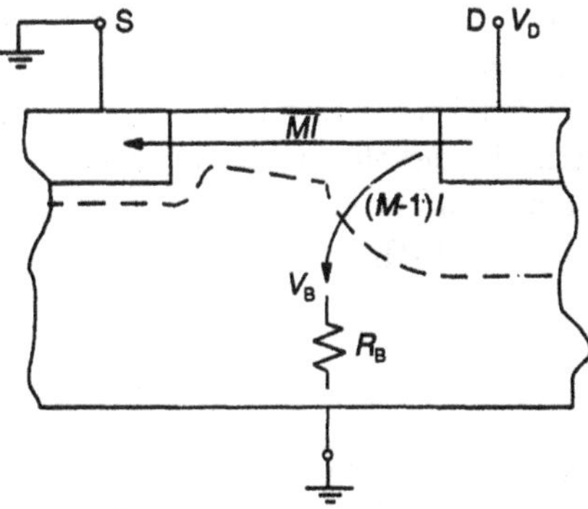

Figure 3.22 Multiplication of the channel and substrate currents due to impact ionization.

Substituting (3.124) for V_B into this equation, solving for I, and using the solution in (3.123), we find the following expression for I_D:

$$I_D = \frac{MI_0}{1 - (M - 1)R_B W v_s C_{ox} K} \quad (3.125)$$

where

$$I_0 \equiv W v_s C_{ox}[V_{GS} - V_{FB} - 2\phi_{FB}(1 + K)]$$

is the drain current predicted when impact ionization is ignored.

According to (3.125), I_D increases linearly with M for the relatively small values of the latter and, hence, of drain voltage. The rate increases with M, and eventually becomes infinitely large for

$$M = 1 + \frac{1}{R_B W v_s K C_{ox}}$$

indicating an avalanche breakdown. This critical value of M and, hence, the breakdown voltage, is a decreasing function of R_B and W but an increasing function of T_{ox} through C_{ox}. In any case, this is a premature breakdown occurring at a finite value of M.

3.5 MOSFET DYNAMICS

To attain a high level of accuracy in MOSFET dynamic analysis, one has to solve partial differential equations of the type described at the beginning of Section 2.4 for a BJT. Except for very high frequency excitations, however, the charge distribution inside the device remains in equilibrium with the time-varying port variables, which enables us to analyze the dynamic MOSFET behavior with quasi-static methods, as we did for a BJT in Section 2.4.1. In this section, we will present a quasistatic analysis, develop modeling equations, and construct equivalent circuits for the dynamic operation of MOSFETs. In-depth coverage of these topics can be found in Tsividis [21].

Modeling the Intrinsic MOSFET

Shown in Figure 3.23 with the rectangle ABCD is the region where the main MOSFET action takes place. The silicon volume of this so-called *intrinsic MOSFET* contains a total electron charge Q_{nt}, which communicates with the source and drain

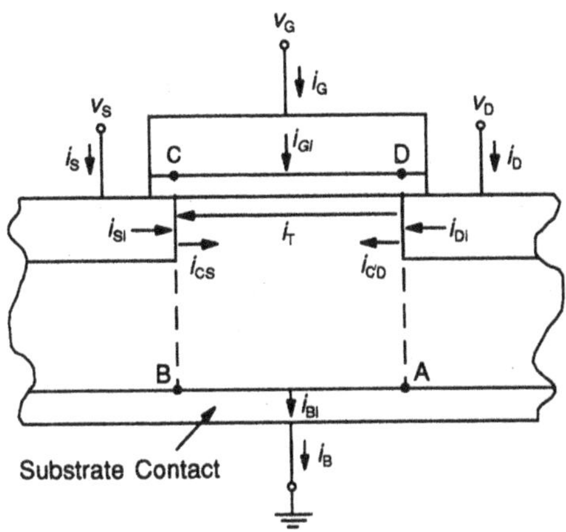

Figure 3.23 Port variables and internal currents in quasistatic modeling of a MOSFET. The intrinsic MOSFET action is confined to the rectangle ABCD.

terminals, and a total hole charge Q_{pt}, which communicates with the substrate terminal. Instantaneous terminal currents are denoted by i_{Si}, i_{Di} and i_{Bi} for source, drain, and substrate, respectively. In the quasistatic treatment, the former two are expressed as

$$i_{Si} = -i_T + i_{CS} \tag{3.126}$$

$$i_{Di} = i_T + i_{CD} \tag{3.127}$$

where the so-called *transport current* i_T is the position-independent channel current, which was previously denoted by I in any of the dc steady-state (static) models. The i_{CS} and i_{CD} terms denote the *charging currents* of the source and drain terminals, which are related to the temporal variation of Q_{nt} by the equation

$$i_{CS} + i_{CD} = \frac{dQ_{nt}}{dt} \tag{3.128}$$

In static operation, both i_{CS} and i_{CD} vanish, hence, $dQ_{nt}/dt = 0$, as required for temporal constancy. The substrate current, which also vanishes in static operation, is related to the variation of Q_{pt} by

$$i_{Bi} = -\frac{dQ_{pt}}{dt} \tag{3.129}$$

Equations (3.126), (3.127), (3.129), and

$$i_{Gi} = i_{Bi} - i_{CS} - i_{CD} \tag{3.130}$$

which is a Kirchhoff's equation written for the gate current, describe all terminal currents for the quasistatic operation of the intrinsic MOSFET. We can turn these into model equations by characterizing Q_{pt} and Q_{nt} in terms of structural parameters and bias voltages, and by properly partitioning dQ_{nt}/dt into source and drain components to identify i_{CS} and i_{CD} separately.

Now suppose that the device operates in nonsaturation, in which the electron and hole charges are localized to the inversion and substrate-bulk regions, respectively; that is,

$$Q_{nt} = -W\int_0^L Q_n\,dy \tag{3.131}$$

$$Q_{pt} = W\int_0^L\int_{x_t}^T qp\,dx\,dy = qWLTN_A - qWN_A\int_0^L x_t\,dy \tag{3.132}$$

where Q_n is the magnitude of the areal charge density in the inversion layer, x_t is the thickness of the depletion region, and T is the thickness of the silicon substrate. In quasistatic operation, the integrands Q_n and x_t are still described by (3.50) and (3.47), respectively. However, these equations are written in terms of V, and not of y. For this reason, we change the variable of integration from y to V by applying the following transformation to (3.131) and (3.132):

$$dy = \frac{W\overline{\mu}_n}{I}Q_n\,dV$$

which is a rearrangement of (3.19). The following expressions are thus obtained:

$$Q_{nt} = -\frac{W^2\overline{\mu}_n}{I}\int_{V_S}^{V_D} Q_n^2\,dV \tag{3.133}$$

$$Q_{pt} = Q_A - qN_A\frac{W^2\overline{\mu}_n}{I}\int_{V_S}^{V_D} x_tQ_n\,dV \tag{3.134}$$

where Q_A is the constant *total acceptor charge*, which appears as the first term on the right-hand side of (3.132). At this point, we introduce the following simplifying assumptions: (1) I can be described with (3.79) of the simplest static model, (2) the threshold voltage appearing in (3.50) of Q_n is independent of V, and (3) x_t can be approximately described with the first two terms of the series expansion of (3.47) around $V = V_S$, that is,

$$x_t \cong \sqrt{\frac{2\epsilon_s}{qN_A}}\left[(V_S + 2\phi_{FB})^{1/2} + \frac{V - V_S}{2(V_S + 2\phi_{FB})^{1/2}}\right]$$

$$= \frac{C_{ox}}{qN_A}[\gamma(V_S + 2\phi_{FB})^{1/2} + \delta'(V - V_S)]$$

where δ' is defined by (3.74). With these simplifications, (3.133) and (3.134) yield

$$Q_{nt} = -\frac{2}{3}\overline{C}_{ox}\frac{(v_{GS} - V_T)^3 - (v_{GD} - V_T)^3}{(v_{GS} - V_T)^2 - (v_{GD} - V_T)^2} \tag{3.135}$$

and

$$Q_{pt} = Q_A - \overline{C}_{ox}\gamma(v_S + 2\phi_{FB})^{1/2} + \frac{1}{3}\overline{C}_{ox}\delta'\frac{2(v_D - v_S)^2 - 3(v_G - v_S - V_T)(v_D - v_S)}{2(v_G - v_S - V_T) - v_D + v_S} \tag{3.136}$$

where $\overline{C}_{ox} \equiv WLC_{ox}$ is the total oxide capacitance.

Equation (3.135) indicates v_{GS} and v_{GD} as the port voltages that control the electron charge. Therefore, the time derivative of Q_{nt} can be expressed as

$$\frac{dQ_{nt}}{dt} = \frac{\partial Q_{nt}}{\partial v_{GS}}\frac{dv_{GS}}{dt} + \frac{\partial Q_{nt}}{\partial v_{GD}}\frac{dv_{GD}}{dt}$$

or alternatively as

$$\frac{dQ_{nt}}{dt} = -C_{GSi}\frac{dv_{GS}}{dt} - C_{GDi}\frac{dv_{GD}}{dt} \tag{3.137}$$

where

$$C_{GSi} \equiv -\frac{\partial Q_{nt}}{\partial v_{GS}} \tag{3.138}$$

$$C_{GDi} \equiv -\frac{\partial Q_{nt}}{\partial v_{GD}} \tag{3.139}$$

are the intrinsic gate-source and gate-drain capacitances. Now, substituting (3.137) into (3.128), and separating the result into source-related and drain-related parts, we obtain

$$i_{CS} = -C_{GSi}\frac{dv_{GS}}{dt} \tag{3.140}$$

and

$$i_{CD} = -C_{GDi}\frac{dv_{GD}}{dt} \tag{3.141}$$

according to which the capacitive charging currents are determined by the time derivatives of the gate-source and gate-drain voltages. The capacitances associated with these currents can be calculated by differentiating (3.135) in accordance with (3.138) and (3.139). The following expressions are obtained.

$$C_{GSi} = \frac{2}{3}C_{ox}\frac{3 - 2\dfrac{V_{DS}}{V_{GS} - V_T}}{\left(2 - \dfrac{V_{DS}}{V_{GS} - V_T}\right)^2} \tag{3.142}$$

$$C_{GDi} = \frac{2}{3}C_{ox}\frac{\left(\dfrac{V_{DS}}{V_{GS} - V_T}\right)^2 - 4\dfrac{V_{DS}}{V_{GS} - V_T} + 3}{\left(2 - \dfrac{V_{DS}}{V_{GS} - V_T}\right)^2} \tag{3.143}$$

Later, we will return to these equations for evaluating C_{GSi} and C_{GDi}. For now, we are going back to (3.129). Considering that Q_{pt} is controlled by all three port voltages v_G, v_S, and v_D, as indicated by (3.136), we can rewrite (3.129) as

$$i_{Bi} = -\frac{\partial Q_{pt}}{\partial v_S}\frac{dv_S}{dt} - \frac{\partial Q_{pt}}{\partial v_D}\frac{dv_D}{dt} - \frac{\partial Q_{pt}}{\partial v_G}\frac{dv_G}{dt}$$

or alternatively as

$$i_{Bi} = C_{SBi}\frac{dv_S}{dt} + C_{DBi}\frac{dv_D}{dt} + C_{GBi}\frac{dv_G}{dt} \tag{3.144}$$

where

$$C_{SBi} \equiv -\frac{\partial Q_{pt}}{\partial v_S} \tag{3.145}$$

$$C_{DBi} \equiv -\frac{\partial Q_{pt}}{\partial v_D} \tag{3.146}$$

and

$$C_{GBi} \equiv -\frac{\partial Q_{pt}}{\partial v_G} \tag{3.147}$$

are the intrinsic source-substrate, drain-substrate, and gate-substrate capacitances, respectively. Differentiating (3.136) in accordance with these definitions, we obtain

$$C_{SBi} = \delta' C_{GSi} \tag{3.148}$$

$$C_{DBi} = \delta' C_{GDi} \tag{3.149}$$

$$C_{GBi} = \frac{1}{3}\overline{C}_{ox}\delta' \frac{\left(\dfrac{V_{DS}}{V_{GS} - V_T}\right)^2}{\left(2 - \dfrac{V_{DS}}{V_{GS} - V_T}\right)^2} \tag{3.150}$$

The quasistatic model of the intrinsic MOSFET, as described by (3.126), (3.127), (3.140), (3.141), and (3.144) can be represented by the equivalent circuit given in Figure 3.24. The circuit comprises a current source representing the transport current i_T and five bias-dependent capacitances. Figure 3.25 shows the variation of these capacitances with V_{DS}. Notice that, for $V_{DS} = 0$, the total oxide capacitance is partitioned equally between the gate-source and gate-drain ports, as one would expect from the fact that the inversion layer is uniform along y for

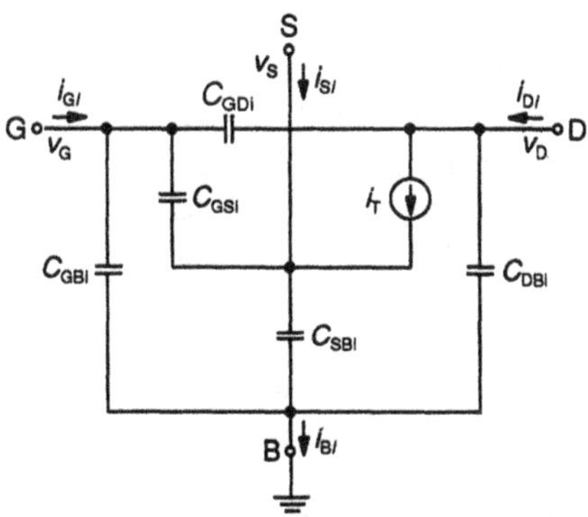

Figure 3.24 An equivalent circuit representing the quasistatic model of the intrinsic MOSFET action.

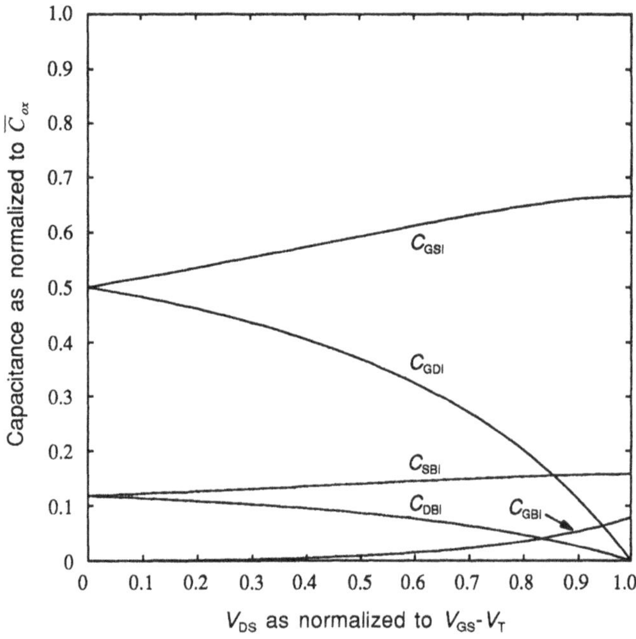

Figure 3.25 The predicted variation of intrinsic MOSFET capacitances with drain-source voltage.

$V_{DS} = 0$. This uniformity is upset when V_{DS} is increased. The charge on the drain side is weakened, which causes C_{GDi} to decrease and C_{GSi} to increase. When the device saturates for $V_{DS} = V_{GS} - V_T$, C_{GD} vanishes because the gate-induced channel disconnects from the drain due to pinch-off. At this point, C_{GSi} assumes two-thirds of the oxide capacitance. Also note that, as the increasing of V_{DS} weakens the channel at the drain side, more of the gate field can penetrate the substrate giving rise to a gate-substrate capacitance C_{GBi}. The source-substrate and drain-substrate capacitances, C_{SBi} and C_{DBi}, are associated with the depletion region beneath the channel. This is why these capacitances are dependent on the substrate doping concentration via δ'. Their dependence on V_{DS} follows the same form as C_{GSi} and C_{GDi} but since δ' is considerably smaller than unity, they are usually much smaller than these two capacitances.

The foregoing discussion of the intrinsic capacitances is based on a nonsaturated operation. As V_{DS} is increased beyond $V_{GS} - V_T$, the intrinsic device saturates but the channel and the underlying depletion region retain their shapes provided that secondary effects, such as channel-length modulation, are negligible. For this reason, all five capacitances become invariant to V_{DS} in saturation.

When the intrinsic device is cut off for $V_{GS} - V_T < 0$, C_{GSi}, C_{GDi}, C_{SBi}, and C_{DBi} are reduced to zero because no channel is left to terminate the gate field. The latter now penetrates the substrate via surface depletion region. This results in an enhanced C_{GBi}. It is left to the reader to show from (3.147), (3.132), and (3.46) that C_{GBi} increases with the decreasing V_G, and reaches $\overline{C}_{ox}$ as the device enters accumulation.

Extrinsic Components and the Complete Model

The intrinsic MOSFET region defined in Figure 3.23 excludes the source and drain junctions and the overlap areas between the gate and two junction surfaces. Due to these extrinsic features, four more capacitances are added to the intrinsic MOSFET equivalent circuit to complete the model for an actual device. The resulting equivalent circuit is shown in Figure 3.26(a). Two of these additional capacitances are the source-substrate and drain-substrate junction capacitances, C_{jS} and C_{jD}, whose bias dependence can be modeled with (2.177), in which v stands for either V_S or V_D. The remaining two capacitances are those of the gate-source and gate-drain overlap areas, and are given by

$$C_{GSo} = C_{GDo} = Wx_{jL}C_{ox}$$

where x_{jL} stands for the length of the overlap area along y.

A Small-Signal Model and Equivalent Circuit

The transport current i_T, as described by the static models, is generally a function of v_{GS}, v_{DS}, and also of v_S through V_T, that is,

$$i_T = f(v_{GS}, v_{DS}, v_S)$$

In the case of a small-signal excitation, the small-signal components of these port variables are interrelated by

$$i_t \equiv di_T = g_m v_{gs} + g_d v_{ds} - g_b v_s \tag{3.151}$$

where

$$g_m \equiv \frac{\partial i_T}{\partial v_{GS}} \tag{3.152}$$

is the *gate transconductance*, or simply the *transconductance*;

$$g_d \equiv \frac{\partial i_T}{\partial v_{DS}} \tag{3.153}$$

is the *drain conductance*; and

$$g_b \equiv -\frac{\partial i_T}{\partial v_S} \tag{3.154}$$

is the *bulk transconductance*. Replacing the current source i_T of the equivalent circuit of Figure 3.26(a) with the two voltage-dependent current sources and one resistance indicated by (3.151), and combining the parallel capacitances, we obtain the small-signal MOSFET equivalent circuit shown in Figure 3.26(b). The capa-

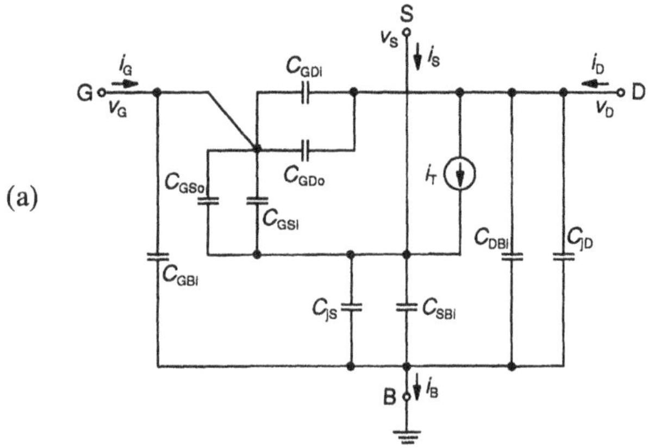

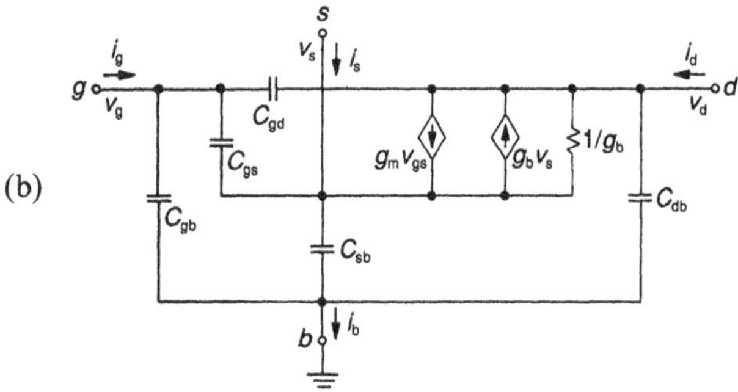

Figure 3.26 (a) The quasistatic equivalent circuit including extrinsic capacitances. (b) A small-signal equivalent circuit.

citive components are still described by the equations already presented in this section. The three additional parameters, g_m, g_d, and g_b can be calculated from (3.152), (3.153), and (3.154) for any desired model equation of i_T. For example, using the intermediate model equation (3.78) for a MOSFET in saturation, and representing the channel-length modulation effect with the parameter λ, we can express i_T as

$$i_T = \frac{K_P}{2}\frac{W}{L}\frac{[v_{GS} - V_T(v_S)]^2}{(1 + \delta')(1 - \lambda v_{DS})}$$

which yields

$$g_m = K_P\frac{W}{L}\frac{V_{GS} - V_T}{(1 + \delta')(1 - \lambda V_{DS})} = \sqrt{2K_P\frac{W}{L}\frac{I}{(1 + \delta')(1 - \lambda V_{DS})}} \tag{3.155}$$

$$g_d = \frac{K_P}{2}\frac{W}{L}\lambda\frac{(V_{GS} - V_T)^2}{(1 - \lambda V_{DS})^2} = \frac{\lambda}{1 - \lambda V_{DS}}I$$

$$g_b = \delta' g_m$$

A figure of merit frequently used for quantifying the speed of a MOSFET is the so-called *intrinsic cutoff frequency* f_T . It is defined as the frequency of a small-signal gate excitation for which the current gain i_d/i_g of a MOSFET with ac-grounded source, bulk, and drain terminals becomes unity. Using the equivalent circuit of Figure 3.26(b), keeping in mind that C_{gd} is zero in saturation, we obtain

$$f_T = \frac{g_m}{2\pi(C_{gb} + C_{gs})}$$

Substituting (3.155) for g_m, $2C_{ox}WL/3$ for C_{gs} [see (3.142)] and $C_{ox}WL\delta'/3$ for C_{gb} [see (3.150)], we turn this equation into

$$f_T = \frac{3\bar{\mu}_n(V_{GS} - V_T)}{2\pi L^2(1 + \delta')(2 + \delta')(1 - \lambda V_{DS})}$$

which clearly shows the importance of minimizing the channel length L. Also obvious is the proportionality of the speed to the effective mobility, which explains why an n-channel MOSFET is about two to three times faster than a p-channel MOSFET of identical structure and bias conditions.

PROBLEMS

3.1 An n-channel MOSFET is specified with $\phi_{FG} = -0.6\text{V}$, $N_A = 1 \times 10^{16}\ \text{cm}^{-3}$, $T_{ox} = 50$ nm, and $Q_{ss} = 8.7 \times 10^{-9}\ \text{C/cm}^2$. Suppose that the source, drain, and substrate terminals are biased with respect to an external reference with 0, 5, and -2V, respectively.

(a) Calculate V_T and V'_T from (3.56) and (3.57). (*Answer:* $V_T = 1$V, $V'_T = 1.95$V.)

(b) Show that the conditions of operation modes are described as

$$V_{GS} < 1\text{V} \quad \text{cutoff}$$

$$1\text{V} < V_{GS} < 6.95\text{V} \quad \text{saturation}$$

$$6.95\text{V} < V_{GS} \quad \text{nonsaturation}$$

by the accurate model; as

$$V_{GS} < 1\text{V} \quad \text{cutoff}$$

$$1\text{V} < V_{GS} < 7.27\text{V} \quad \text{saturation}$$

$$7.27\text{V} < V_{GS} \quad \text{nonsaturation}$$

by the intermediate model; and as

$$V_{GS} < 1\text{V} \quad \text{cutoff}$$

$$1\text{V} < V_{GS} < 6\text{V} \quad \text{saturation}$$

$$6\text{V} < V_{GS} \quad \text{nonsaturation}$$

by the simple model.

(c) Assuming $L' = L$, $W/L = 10$, and $\overline{\mu}_n = 600$ cm^2/V · s, plot the MOSFET transfer characteristic $I = f(V_{GS})$ for $0 \leq V_{GS} \leq 10$V using all three models separately.

(d) Calculate the transconductance $g_m \equiv \partial I/\partial V_{GS}$ for $V_{GS} = 5$V using all three models separately. Compare the results.

(e) Calculate the transconductance using the simple model but taking into consideration the transversal-field dependence of $\overline{\mu}_n$ as described by (3.93). You can use $U_c = 10^4$ V/cm, $U_e = 0.1$, $U_t = 0.5$, and $\overline{\mu}_o = 600$ cm^2/V · s. Plot the results as in part (c) and compare them with the corresponding curves derived for field-independent mobility.

3.2 In Section 3.2.2, we constructed a transversal energy-band diagram on the basis of flat bands in the polysilicon gate. The argument was made that a very heavy doping concentration precluded any significant bending in that part of the MOSFET structure. The assumption of very heavy gate doping is indeed justifiable when doping is implemented with the conventional technique of chemical vapor deposition of $POCl_3$, but may not be justifiable if ion implantation is used for doping the gate [22]. In this exercise, you are expected to analyze the effect of gate band-bending on the surface field $\mathscr{E}(0)$ and, hence, on the channel charge density and current. Shown in Figure P3.2 is a transversal energy-band diagram, on which ψ_{SG} denotes band-bending in the gate. You can treat the gate as an "n-type substrate" like that of a p-channel MOSFET; therefore, regard ψ_{SG} as the surface potential of this substrate.

(a) Show that band-bending in the gate modifies (3.29) as

$$V_G = V_{FB} + \psi_s - \psi_{SG} + \frac{\epsilon_s}{C_{ox}}\mathscr{E}(0)$$

and that the boundary field $\mathscr{E}_G(0)$ of the gate, as defined in Figure P3.2, is related to $\mathscr{E}(0)$ by

$$\mathscr{E}_G(0) = \frac{Q_{SG} + Q_{ss}}{\epsilon_s} - \mathscr{E}(0) \qquad \text{(P3.2.1)}$$

where Q_{SG} is an oxide fixed charge located at the gate-SiO_2 interface.

(b) You know that for a p-type substrate to be strongly inverted, the condition

$$\mathscr{E}(0) > \sqrt{\frac{2qN_A}{\epsilon_s}(V + 2\phi_{FB})}$$

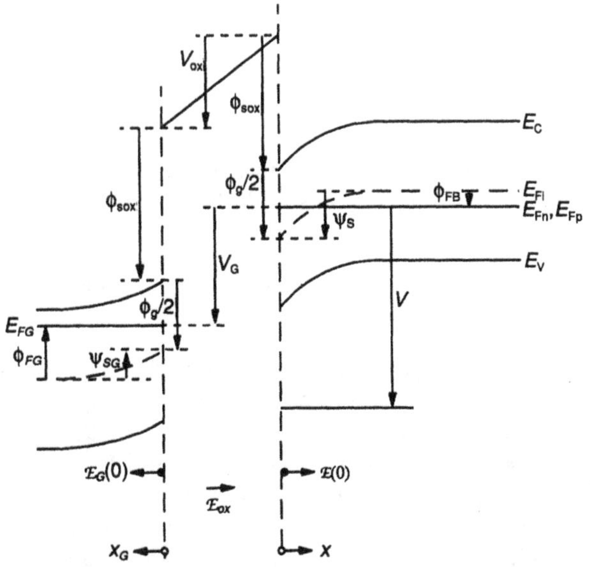

Figure P3.2

must be satisfied. Ignoring $Q_{SG} + Q_{ss}$ in (P3.2.1), persuade yourself that the space-charge region inside the n-type gate can be depleted or inverted but not accumulated when a strong inversion condition prevails in the p-type monocrystalline substrate. Now show that the gate is depleted for

$$\sqrt{\frac{2qN_A}{\epsilon_s}(V + 2\phi_{FB})} < \mathscr{E}(0) < \sqrt{\frac{2qN_G}{\epsilon_s}(-2\phi_{FB})} \qquad \text{(P3.2.2)}$$

and is inverted for

$$\sqrt{\frac{2qN_G}{\epsilon_s}(-2\phi_{FG})} < \mathscr{E}(0) \qquad \text{(P3.2.3)}$$

where N_G is the gate doping concentration and ϕ_{FG} is the gate Fermi potential.

(c) Since N_G is usually many orders of magnitude larger than N_A, the condition stated by (P3.2.3) is unlikely to occur. Therefore, the gate is usually depleted when the substrate is inverted. Based on this conclusion, show that

$$\psi_{SG} = -\frac{\epsilon_s}{2qN_G}\mathscr{E}^2(0)$$

and

$$\mathscr{E}(0) = \frac{qN_G}{C_{ox}}\left[\sqrt{1 + \frac{2C_{ox}^2}{q\epsilon_s N_G}(V_G - V_{FB} - 2\phi_{FB} - V)} - 1\right] \qquad \text{(P3.2.4)}$$

(d) Supposing that the second term under the square root sign in (P3.2.4) is much smaller than unity, you should be able to obtain

$$\mathscr{E}(0) \cong \frac{C_{ox}}{\epsilon_s}(V_G - V_{FB} - 2\phi_{FB} - V)\left[1 - \frac{C_{ox}^2}{2q\epsilon_s N_G}(V_G - V_{FB} - 2\phi_{FB} - V)\right]$$

Now, calculate the percentage reduction caused by the gate-depletion effect in the surface field (with respect to the surface field predicted by the primary model) at the source side of the channel in a MOSFET of T_{ox} = 9.8 nm, N_A = 1 × 10^{16} cm^{-3}, $Q_{ss} \cong 0$, N_G = 1.2 x 10^{19} cm^{-3}, V_G = 5V, and V_S = 0. (*Answer:* 16%.)

3.3 The substrate doping concentration employed in a n-well CMOS technology is 5 x 10^{15} cm^{-3} for the n-channel MOSFET and 3 x 10^{16} cm^{-3} for the p-channel MOSFET. Suppose that both transistors have the same gate oxide of thickness 20 nm, the same oxide fixed charge of density 2 x 10^{-8} C/cm^2, and the same n-type poly-silicon gate of Fermi potential of -0.6V.

(a) Calculate the zero-bias threshold voltages of these two devices. (*Answer:* $V_{T0} = 0.187$V for the n-channel device and $V_{T0} = -1.6$V for the p-channel device.)

(b) Obviously, the threshold voltages calculated in part (a) do not satisfy the usual CMOS specifications of $0.5\text{V} < V_{T0} < 1\text{V}$ for the n-channel device and $-1\text{V} < V_{T0} < -0.5\text{V}$ for the p-channel device. Show that a single p-type threshold-adjustment implantation of dose 1.05 x 10^{12} cm^{-2} and of effective adjustment depth $d_t = 30$ nm can shift these threshold voltages to 0.75V for the n-channel device and -0.74V for the p-channel device, and thus can satisfy the specifications. (You can use $N_A = D_T/d_t$ to calculate the effective doping concentration in the buried channel of the p-channel device.)

3.4 The flatband voltage, as described by (3.30), depends on the gate-oxide thickness through C_{ox} only. This is true even in the case of a metal-gate MOSFET, for which the only change in (3.30) is the replacement of $\phi_{FG} - \phi_{FB}$ by some other T_{ox}-independent quantity. Relying on this fact, one can extract the value of Q_{ss} by applying the following experimental procedure: (1) Grow the oxide for which Q_{ss} is to be characterized. (2) Measure the thickness of the oxide. (3) Place a "mercury" probe on top of the oxide as a "portable" gate material. (4) Measure V_{FB}. (5) Remove the mercury probe. (6) Etch back a thin layer of oxide. (7) Repeat steps 2 through 6 several times. (8) Plot V_{FB} as a function of T_{ox} and extract Q_{ss} form the slope of this plot.

(a) Convince yourself that, theoretically, this procedure is a viable one.

(b) A closely related experimental technique exists by which one can characterize the charges created by radiation inside the gate oxide. You will learn more about this technique in part (c) of this exercise. For now, however, suppose that a sheet charge of density Q_{ox} [C/cm^2] resides inside the oxide at a distance D from the oxide-semiconductor interface. Find an expression for the flatband voltage in terms of Q_{ox}, D, Q_{ss}, T_{ox}, ϵ_{ox}, ϕ_{FG}, and ϕ_{FB}, and compare it with (3.30).

(c) The first four steps of the experimental technique by which an oxide charge is characterized are identical to those of the Q_{ss} extraction procedure described in this problem. Suppose that the flatband voltage measured at the end of step 4 is V_{FB0}. After removing the mercury probe, the oxide is irradiated to create an oxide charge. This is followed by several etching and V_{FB} measurement steps. Suppose that, after the nth etching, the oxide thickness is T_{oxn} and the flatband voltage is V_{FBn}. Show that, by using a plot of $\Delta V_{FB} \equiv V_{FBn} - V_{FB0}$ versus T_{oxn}, you can extract both Q_{ox} and its centroid D, provided that you know the value of Q_{ss}. You can refer to the paper by Ramesh et al. [23] for a recent application of this technique, and to the references cited therein for earlier work.

3.5 Up to now, we have treated the width W of a MOSFET channel as a structural constant. Experimental observations, however, indicate that, effectively, the width is an increasing function of the gate voltage. Chia and Hu [24] attribute this effect to a lateral spreading of inversion beyond the designed width of the channel region. For an explanation of their argument, first note that the local oxidation technique employed for selectively growing the field oxide of a MOSFET yields a gradual change in the oxide thickness between the main channel region and the field region, as shown in Figure P3.5. The resulting tapered oxide region, called a *bird's beak*, is intentionally overlaid with an extension of the polysilicon gate in order to ensure full coverage of the main channel region even if a worst case misalignment occurs during the lithography of the poly layer. The threshold voltage in the bird's beak area is an increasing function of the position variable z because the oxide thickness increases with the latter. As we increase the gate voltage, an increasingly larger silicon area will be inverted beneath the bird's beak, thus causing an outspreading of the channel beyond the main channel area.

(a) Convince yourself that the effect just described can be included in the current-voltage characteristic equations by replacing W with an "effective width"

$$W_{\text{eff}} \equiv W + \frac{2}{Q_n(0)} \int_0^{z_L} Q_n(z)\,dz$$

where W is the width of the main channel area, $Q_n(0)$ is the channel charge density in the main channel area, $Q_n(z)$ is the channel charge density under the bird's beak, and z_L is the lateral extension of the inversion layer.

(b) Assuming a linear spatial variation $T_{ox} = az + T_{ox0}$ (a is a constant) for the oxide thickness of the bird's beak region, and a uniformly doped substrate[1], show that

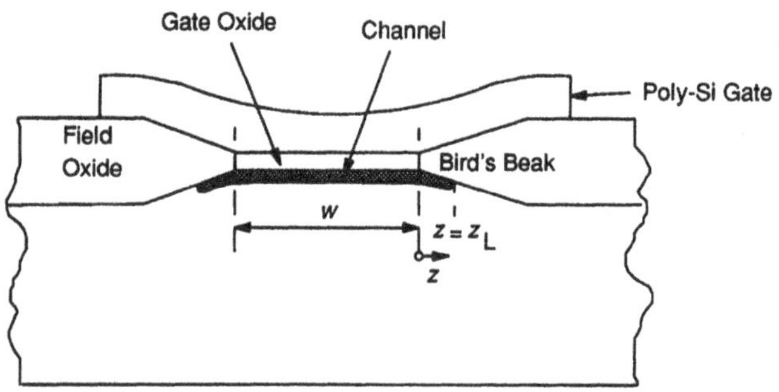

Figure P3.5

[1]Usually, the field region is heavily doped with boron to increase the threshold voltage. This so-called *channel-stopper* doping, which also enhances the doping concentration beneath the bird's beak region, is ingnored in this exercise.

$$z_L = \frac{V_G - V - V_{T0}(V)}{a\gamma_0\sqrt{V + 2\phi_{FB}}} T_{ox0}$$

where V is the channel potential, $V_{T0}(V)$, γ_0, and T_{ox0} are, respectively, the threshold voltage, body factor, and oxide thickness of the main channel region.

(c) Now, show that the deviation $\Delta W \equiv W_{\text{eff}} - W$ can be described with the equation

$$\Delta W = \frac{T_{ox0}}{a}\left\{\left[1 + \frac{\gamma_0\sqrt{V + 2\phi_{FB}}}{V_G - V - V_{T0}(V)}\right] \ln\left[1 + \frac{V_G - V - V_{T0}(V)}{\gamma_0\sqrt{V + 2\phi_{FB}}}\right] - 1\right\}$$

(d) Plot ΔW as a function of gate overdrive $V_G - V_{T0}$ for the source edge of the channel assuming $V_S = 0$, $T_{ox0} = 40$ nm, $a = 1.2$, and $N_A = 1 \times 10^{16}$ cm^{-3}.

3.6 The silicon-on-insulator "DELTA" structure developed by Hisamoto et al. [25] is essentially a dual-channel MOSFET built on a very thin substrate. A simplified schematic of this structure is given in Figure P3.6(a). The thickness T of the substrate is so small that when the SiO_2-Si interfaces located at $x = 0$ and $x = T$ are inverted, the rest of the substrate is already fully depleted. Also note that the substrate has no terminal available for an external definition of its potential. This is why the three terminal voltages V_G, V_D, and V_S are defined with respect to an arbitrary reference level on the transversal energy-band diagram shown in Figure P3.6(b).

(a) The threshold voltage at an arbitrary lateral location y is defined as

$$V_T \equiv V_G - V(y)\,|_{\phi_{Fn}(0) = -\phi_{FB}}$$

where $\phi_{Fn}(0)$ is the electron quasi-Fermi potential at $x = 0$ (and, of course, at $x = T$) and $V(y)$ is the channel potential. Show that

$$V_T = \phi_{FG} + \phi_{FB} - \frac{Q_{ss}}{C_{ox}} + \frac{qN_AT}{2C_{ox}}$$

Discuss the difference between this result and (3.42).

(b) Following the procedure of Section 3.2.3, develop an accurate strong inversion model $I = f(V_{GS}, V_{DS})$ for this structure and compare the result with the three strong inversion models developed for the conventional MOSFET.

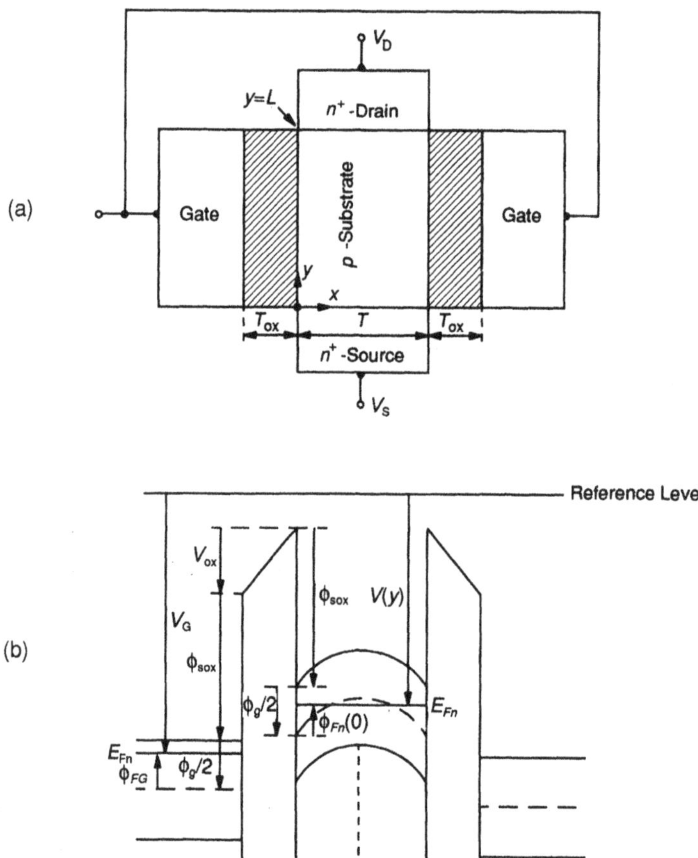

Figure P3.6

3.7 Lightly doped drain (LDD) MOSFETs are designed to reduce the lateral field in the pinch-off region for the purpose of minimizing device degradation due to energetic carrier injection into the gate oxide. As shown in Figure P3.7,

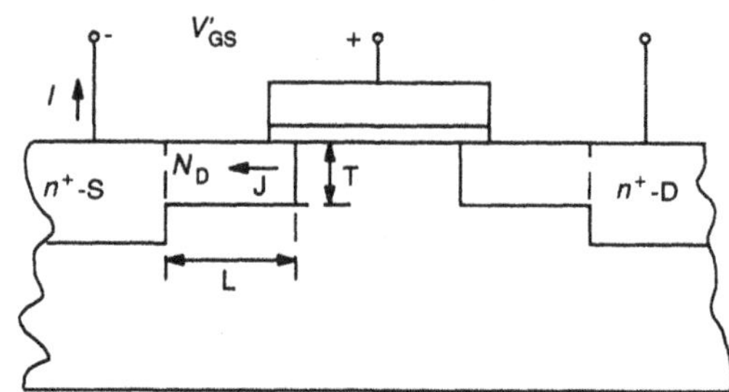

Figure P3.7

these devices have relatively lightly doped source and drain extensions into the channel.

(a) Assuming that current conduction in these lightly doped regions (doped uniformly to a concentration N_D) is due to an electron drift with a field-dependent mobility,

$$\mu = \frac{v_s}{\mathscr{E} + \mathscr{E}_c}$$

and that the density J of this current is independent of the transversal position along the entire thickness T, show that the potential drop along each region can be expressed as

$$\psi = IR$$

where

$$R = \frac{R_{S0}}{1 - (R_{S0} I / \mathscr{E}_c L)}$$

is the current-dependent resistance of the region,

$$R_{S0} \equiv \frac{L}{TWq\mu_{no}N_D}$$

is the zero-current resistance, and

$$\mu_{n0} = \frac{v_s}{\mathscr{E}_c}$$

is the zero-field mobility [26].

(b) Due to the presence of R, the externally applied gate-source voltage V'_{GS} is related to the "intrinsic" V_{GS} by

$$V'_{GS} = V_{GS} + \psi$$

Assuming velocity saturation in the actual channel under high gate-bias conditions, find an expression for the "apparent" transconductance $g_m \equiv \partial I / \partial V'_{GS}$ in terms of $\mathscr{E}_c$, L, R_{S0}, I, and the intrinsic transconductance $g_{mo} \equiv W v_s C_{ox}$. Compare your result with the transconductance of a conventional MOSFET operating under similar conditions.

3.8 Down-scaling of MOSFET dimensions has been instrumental in the evolution of density and performance of integrated circuits for the past two decades.

(a) Suppose that T_{ox}, L, and W, as well as the source-drain junction dimensions, are scaled down by a common factor $S > 1$; that is, all of these dimensions are multiplied by $1/S$, but the doping concentrations and bias voltages are left unchanged, and the threshold voltage is adjusted to remain the same. Show that the main parameters and variables would change by the following:

- Integration density $[(\propto(WL)^{-1}]$: S^2
- Body factor: $1/S$
- Transconductance parameter (K_P): S
- Channel-length modulation parameter: S
- Punch-through voltage: $1/S^2$
- Channel current: S for low-field, 1 for velocity saturation
- Static power dissipation ($I \times V$): S for low-field, 1 for velocity saturation
- Dynamic power dissipation ($\propto WLC_{ox}V^2$): $1/S$
- Static power density per unit gate area: S^3 or S^2
- Dynamic power density per unit gate area: S
- Transconductance (g_m): S or 1
- Cutoff frequency: S^2 or S
- Field in gate oxide: S
- Peak field in source or drain junctions: 1

(b) The scaling rule defined in part (a) improves integration density and speed. However, the channel-length modulation and punch-through effects are also strengthened, which would eventually set a lower limit on the scaled dimensions. Also obvious are the limitations imposed by power density and oxide field. A scaling rule, called *constant-field scaling*, was proposed by Dennard et al. [27] in 1974 for the purpose of overcoming these limitations. It prescribes the scaling factor $1/S$ not only for dimensional scaling but also for bias-voltage scaling. Furthermore, the doping concentrations are also supposed to be scaled (up) by S. Assuming an unscaled gate overdrive ($V_{GS} - V_T$), find the scale factors of the parameters and variables of part (a) for constant-field scaling and compare the results.

(c) The constant-field scaling has not been applied as proposed because the scaling of bias voltages has been incompatible with the TTL-defined supply voltage standard of 5V. Repeat part (b) assuming constant bias voltages, and compare the results with those of parts (a) and (b).

3.9 The subthreshold behavior of a MOSFET is usually characterized with the so-called *subthreshold-swing* parameter

$$S \equiv \left(\frac{\partial \log I}{\partial V_{GS}}\right)^{-1}$$

(a) Assuming $V_{TX} - V_{FB} \gg kT/q$, show that S can be expressed as

$$S = \frac{kT}{q} \frac{2.3}{1 - (dV_{TX}/dV_{GS})}$$

(b) Oftentimes S is expressed as

$$S = 2.3\frac{kT}{q}\left(1 + \frac{C_t}{C_{ox}}\right)$$

where $C_t \equiv \epsilon_s/x_t$ is the capacitance of the surface depletion region of width x_t. Show that this physically more informative expression is equivalent to the one given in part (a).

(c) The smaller the value of S, the higher the control of the gate on the subthreshold current. A small S is highly desirable when the MOSFET is used as a switch because any residual subthreshold current in the off state of the device represents a leakage path. Which structural parameters are effective on S? How do the scaling rules defined in Problem 3.8 affect S?

3.10 Suppose that the n-channel MOSFET shown in Figure P3.10(a) is an "intrinsic" device as defined in the quasistatic model. Assuming $V_T = 1\text{V}$, $N_A = 5.6 \times 10^{15}\ \text{cm}^{-3}$, $T_{ox} = 40\ \text{nm}$, $W = 10\ \mu\text{m}$, $L = 2\ \mu\text{m}$, and using the simple strong inversion model for the transport current, determine the waveforms $i_{Bi}(t)$, $i_{Di}(t)$, $i_{Si}(t)$, and $i_{Gi}(t)$ for the gate-voltage drive specified in Figure P3.10(b). Note that the reference direction selected for i_{Si} in this problem is opposite to that of the text.

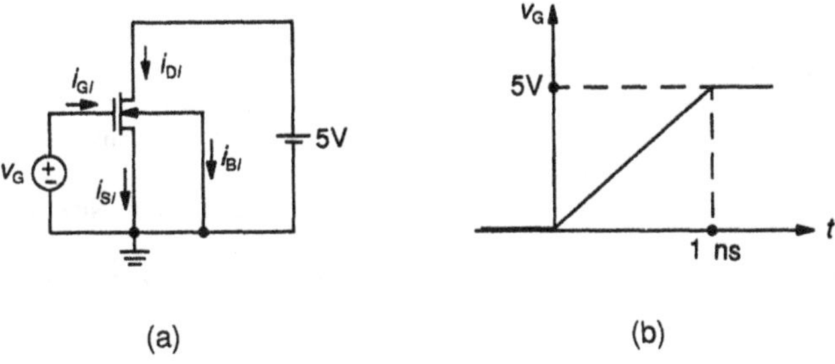

Figure P3.10

REFERENCES

[1] Chen, John Y., *CMOS Devices and Technology for VLSI*, Englewood Cliffs, NJ: Prentice Hall, 1990.
[2] Tsividis, Yannis P., *Operation and Modeling of the MOS Transistor*, New York: McGraw-Hill, 1987.
[3] Meyer, J. E., "MOS Models and Circuit Simulation," *RCA Review*, vol. 32, 1971, p. 42.
[4] Merckel, G., J. Borel, and N. Z., Cupcea, "An Accurate Large-signal MOS Transistor Model for Use in Computer-aided Design," *IEEE Trans. on Electron Devices*, Vol. ED-19, 1972, p. 681.
[5] Shichman, H., and D.A., Hodges, "Modeling and Simulation of Insulated-gate Field-effect Transistor Switching Circuits," *IEEE J. of Solid-State Circuits*, Vol. SC-3, 1968, p. 285.
[6] Barron, M. B., "Low-level Currents in Insulated-gate Field-effect Transistors," *Solid-State Electron.*, Vol. 15, 1972, p. 293.
[7] Troutman, R. R., "Subthreshold Design Considerations for Insulated Gate Field-effect Transistors," *IEEE J. of Solid-State Circuits*, Vol. SC-9, 1974, p. 55.
[8] Troutman, R. R., "Subthreshold Slope for Insulated Gate Field Effect Transistors," *IEEE Trans. on Electron Devices*, Vol. ED-22, 1975, p. 1049.
[9] Van Overstraeten, R. J., G. J., Declerck, and P. A., Muls, "Theory of MOS Transistor in Weak Inversion—New Method to Determine the Number of Surface States," *IEE Trans. on Electron Devices*, Vol. ED-22, 1975, p. 282.
[10] Brews, J. R., "Subthreshold Behavior of Uniformly and Nonuniformly Doped Long-channel MOSFET," *IEEE Trans. on Electron Devices*, Vol. ED-26, 1979, p. 1282.
[11] Antognetti, P., and G., Massobrio, *Semiconductor Device Modeling with SPICE*, New York: McGraw-Hill, 1988, p. 164.
[12] Rideout, V. L., F. H. Gaensslen, and A., LeBlanc, "Device Design Considerations for Ion-implanted N-channel MOSFETs," *IBM J. Res. Develop.*, Vol. 19, 1975, p. 50.
[13] Van Der Tol, J. Michael, and Savvas G., Chamberlain, "Potential and Electron Distribution Model for the Buried-channel MOSFET," *IEEE Trans. on Electronic Devices*, Vol. 36, 1989, p. 670.
[14] Sodini, C. G., P.-K. Ko, and J. L., Moll, "The Effect of High Fields on MOS Device and Circuit Performance," *IEEE Trans. on Electron Devices*, Vol. ED-31, 1984, p. 1386.
[15] Tsividis, Yannis P., *Operation and Modeling of the MOS Transistor*, New York: McGraw-Hill, 1987, p. 171.
[16] Antognetti, P., and G., Massobrio, *Semiconductor Device Modeling with SPICE*, New York: McGraw-Hill, 1988, p. 163.
[17] Barnes, John J., Katsuhiro Shimohigashi, and Robert W. Dutton, "Short-channel MOSFET's in the Punchthrough Current Mode," *IEEE Trans. on Electron Devices*, Vol. ED-26, 1979, p. 446.
[18] Yau, L. D., "A Simple Theory to Predict the Threshold Voltage of Short-channel IGFETs," *Solid-State Electron.*, Vol. 17, 1974, p. 1059.
[19] Akers, L. A., and J. J., Sanchez, "Threshold Voltage Models of Short, Narrow and Small Geometry MOSFET's: A Review," *Solid-State Electron.*, Vol. 25, 1982, p. 621.
[20] Toyabe, Toru, and Shojiro, Asai, "Analytical Models of Threshold Voltage and Breakdown Voltage of Short-channel MOSFET's Derived from Two-dimensional Analysis," *IEEE Trans. on Electron Devices*, Vol. ED-26, 1979, p. 453.
[21] Tsividis, Yannis P., *Operation and Modeling of the MOS Transistor*, New York: McGraw-Hill, 1987, Chaps. 7 and 8.
[22] Habas, Predrag, and John V., Faricelli, "Investigation of the Physical Modeling of the Gate-Depletion Effect," *IEEE Trans. Electron Devices*, Vol. 39, 1992, p. 1496 and the references cited therein.
[23] Ramesh, K., A., Agarwal, A. N. Chandorkar, and J., Vasi, "Role of Electron Traps in the Radiation Hardness of Thermally Nitrided Silicon Dioxide," *IEEE Electronic Device Lett.*, Vol. 12, 1991, p. 658.

[24] Chia, Y.-T., and J., Hu, "A Method to Extract Gate-Bias-Dependent MOSFET's Effective Channel Width," *IEEE Trans. on Electron Devices*, Vol. 38, 1991, p. 424.
[25] Hisamoto, D., T. Kaga, and E., Takeda, "Impact of the Vertical SOI 'DELTA' Structure on Planar Device Technology," *IEEE Trans. on Electron Devices*, Vol. 38, 1991, p. 1419.
[26] Reich, R. K., D.-H. Ju, and A. M., Sekela, "Velocity Saturation Limitations of Lightly Doped Drain Transistors," *IEEE Trans. on Electron Devices*, Vol. 35, 1988, p. 444.
[27] Dennard, R. H., F. H., Gaensslen, H. N., Yu, V. L., Rideout, E. Bassous, and A., LeBlanc, "Design of Ion Implanted MOSFETs with Very Small Physical Dimensions," *IEEE J. Solid-State Circuits*, Vol. SC-9, 1974, p. 256.

Index